Jan. 31, 1996

To Mino,

With fond memories of London, 1990.

Bob

OPTICAL SOURCES, DETECTORS, AND SYSTEMS

OPTICS AND PHOTONICS
(formerly Quantum Electronics)

EDITED BY

PAUL F. LIAO

Bell Communications Research, Inc.
Red Bank, New Jersey

PAUL L. KELLEY

Lincoln Laboratory
Massachusetts Institute of Technology
Lexington, Massachusetts

IVAN KAMINOW

AT&T Bell Laboratories
Holmdel, New Jersey

A complete list of titles in this series appears at the end of this volume.

OPTICAL SOURCES, DETECTORS, AND SYSTEMS

FUNDAMENTALS AND APPLICATIONS

Robert H. Kingston

Department of Electrical Engineering and Computer Science
Massachusetts Institute of Technology

ACADEMIC PRESS

San Diego Boston New York
London Sydney Tokyo Toronto

This book is printed on acid-free paper. ♾

Academic Press, Inc.
A Division of Harcourt Brace & Company
525 B Street, Suite 1900, San Diego, California 92101-4495

United Kingdom Edition published by
Academic Press Limited
24-28 Oval Road, London NW1 7DX

Library of Congress Cataloging-in-Publication Data

Kingston, Robert Hildreth, date.
Optical sources, detectors, and systems : fundamentals and applications / by Robert H. Kingston.
p. cm. — (Optics and Photonics series)
Includes index.
ISBN 0-12-408655-1 (case : alk. paper)
1. Photonics. 2. Optical detectors. 3. Optical communications. 4. Imaging systems. I. Title. II. Series.
TA1520.K56 1995
621.36—dc20 95-12389
CIP

PRINTED IN THE UNITED STATES OF AMERICA
95 96 97 98 99 00 BB 9 8 7 6 5 4 3 2 1

Contents

Preface

During the last decade, optical sources and detectors have become mainstays of our technology as exemplified by fiber optic communications, the camcorder, and the supermarket checkout counter. In this textbook, I have treated the fundamentals of both the sources and the detectors and their performance in such systems. In a sense the main theme of this book is *noise*, the fluctuation or uncertainty in a measured signal voltage which can lead to transmission errors in a fiber optic link or "snow" in a TV image. The noise is an inherent part of the photodetection process and is also produced by an electric circuit form of thermal radiation. These noise processes are treated in detail throughout the text and lead to quantitative predictions and figures of merit for overall system performance.

The required background for the student or general reader is an understanding of probability densities and electrical system concepts such as Fourier transforms and complex notation ($\mathrm{Re}[Ee^{j\omega t}]$), as well as semiconductor device principles. Although helpful, training in quantum mechanics is not essential, because the needed elements are developed in the text.

I am indebted to many colleagues for advice and suggestions, especially Stephen Alexander and Roy Bondurant on electrical and laser amplifiers , William Keicher, Herbert Kleiman and Harold Levenstein on radar systems, and Barry Burke on charge coupled devices (CCDs).

Robert H.Kingston

Symbols

A	Einstein A coefficient	sec^{-1}
B	Einstein B coefficient	$m^3\ J^{-1} sec^{-1}$
c	Velocity of light, 3.00×10^8	$m\ sec^{-1}$
dB	Decibel, $10log_{10}(P_{out}/P_{in})$	
D	Detectivity	W^{-1}
D^*	Specific detectivity	$cm\ \sqrt{Hz}\ W^{-1}$
f	Electric circuit frequency	Hz
F	Noise factor, avalanche photodiode	
$F(x)$	Normalized Planck radiance function	
$G(x)$	Normalized Planck photon count function	
h	Planck's constant, 6.6×10^{-34}	$J\ sec$
H	Radiance	$W\ m^{-2}$
H^ϕ	Photon count radiance	$sec^{-1}\ m^{-2}$
H_λ	Spectral radiance, $dH/d\lambda$	$W\ m^{-3}$
H_ν	Spectral radiance, $dH/d\nu$	$W\ m^{-2}\ Hz^{-1}$
I	Intensity or irradiance	$W\ m^{-2}$
I_ν	Spectral irradiance	$W\ sec\ m^{-2}$
$I(>x)$	Fraction of radiance, H, greater than $x = h\nu/kT$	
$J(>x)$	Fraction of photon radiance, H^ϕ, greater than x	
$K(x)$	Normalized spectral density function	
k	Boltzmann's constant, 1.38×10^{-23}	$J\ K^{-1}$
m_e	Free electron mass, 9.1×10^{-31}	kgm
q	Electron charge, 1.6×10^{-19}	C
$\Re$	Responsivity	$A\ W^{-1}$
u	Energy density	$J\ m^{-3}$
U	Energy	J
α	Absorption or ionization coefficient	m^{-1}
γ	Gain coefficient	m^{-1}
Γ	Noise factor, photomultiplier	
δ	Secondary emission ratio	
δ	Noise current ratio, nonsignal to signal	
ε	Emissivity	
ν	Optical wave frequency	Hz
ρ	Reflectivity	
σ	Stefan-Boltzmann constant, 5.67×10^{-8}	$W\ m^{-2}\ K^{-4}$
σ	Radar cross section	m^2

Chapter 1
Blackbody Radiation, Image Plane Intensity, and Units

Optical and infrared sources are of two general types, either *incoherent* such as thermal radiation or *coherent* such as that from a laser. Here we first treat the classical thermal or *blackbody* radiation emitted by any body at finite temperature. In particular we are most interested in "room temperature" radiation, that from a body at *300 K*, and solar radiation corresponding to the sun's temperature of *5800 K*. Blackbody radiation is incoherent in the sense that there is an infinite set of optical frequencies present and the phase of each constituent frequency term is a random function of the direction of propagation. In contrast, the coherent radiation obtainable from a laser can be treated as essentially monochromatic or single frequency with uniform phase over a plane or spherical wavefront. Actually, the *coherence* of a source is a matter of degree and may be measured in a quantitative manner *(Goodman, 1985; Saleh and Teich, 1991, Ch. 10).* Although we call thermal radiation *incoherent*, if we pass it through a pinhole of diameter less than the radiation wavelength, the resultant spherical wave is *spatially coherent* (but very low in intensity). Similarly, if we pass thermal radiation through a narrowband frequency filter it becomes *temporally partially coherent.*

Following our treatment of blackbody radiation, we derive the expected intensity in an optical image plane, a treatment valid for both a blackbody radiation source and any type of radiation scattered from a diffuse surface. We then discuss numerical values and introduce a convenient set of units and techniques for calculating intensities and powers, concluding with a brief discussion of the *lumen*. The results of this chapter are essential for understanding the detection and measurement of thermal or solar radiation. In addition, when we wish to detect laser radiation, we shall use the results to calculate the effects of the "background" thermal or solar radiation on detection efficiency. We shall also use our theoretical approach as a key element of the treatment of laser action in Chapter 2 and thermally induced *electrical* noise in Chapter 4.

1.1 Planck's Law

By convention and definition *blackbody* radiation describes the intensity and spectral distribution of the optical and infrared power emitted by an ideal *black* or completely absorbing material at a uniform temperature T. The radiation laws are derived by considering a completely enclosed container whose walls are uniformly maintained at temperature T, then calculating the internal energy density and spectral distribution using thermal statistics. Consideration of the equilibrium interaction of the radiation with the chamber walls then leads to a general expression for the emission from a "gray" or "colored" material with nonzero reflectance. The treatment yields not only the spectral but the angular distribution of the emitted radiation.

Although we usually refer to blackbody radiation as "classical", its mathematical formulation is based on the quantum properties of electromagnetic radiation. We call it classical since the form and the general behavior were well known long before the correct physics was available to explain the phenomenon. We derive the formulas using Planck's original hypothesis, and it is in this derivation, known as Planck's law, that the quantum nature of radiation first became apparent. We start by considering a large enclosure containing electromagnetic radiation and calculating the energy density of the contained radiation as a function of the optical frequency ν. To perform this calculation we assume that the radiation is in equilibrium with the walls of the chamber, that there are a calculable number of "modes" or standing-wave resonances of the electromagnetic field, and that the energy per mode is determined by thermal statistics, in particular by the Boltzmann relation

$$p(U) = Ae^{-U/kT} \tag{1.1}$$

where $p(U)$ is the probability of finding a mode with energy, U; k is the Boltzmann constant; T, the absolute temperature; and A is a normalization constant.

Example: The Boltzmann distribution will be used frequently in this text since it has such universal application in thermal statistics. As an interesting example, let us consider the variation of atmospheric pressure with altitude under the assumption of constant temperature. The pressure, at constant temperature, is proportional to the density and thus to the probability of finding an air molecule at the energy U associated with altitude h, given by $U = mgh$, with m the molecular mass and g the acceleration of gravity. Thus the variation of pressure with altitude may be written

$$P(h) = P(0)e^{-mgh/kT}$$

and the atmospheric pressure should drop to $1/e$ or 37% at an altitude of $h = kT/mg$. Using 28 as the molecular weight of nitrogen, the principal constituent, yields

$$mg = 28(1.66 \times 10^{-27})\, 9.8 = 4.5 \times 10^{-25} \textit{ newtons}$$

$$kT = 1.38 \times 10^{-23}(300) = 4.1 \times 10^{-21} \textit{ joules}$$

$$h(37\%) = 9 \times 10^{3} \textit{ meters} = 9 \textit{ km or } 30{,}000 \textit{ feet.}$$

This is quite close to the nominal observed value of 8 *km*, determined by the more complicated true molecular distribution and a significant negative temperature gradient. We discuss a simpler way of calculating energies in section 1.5.

Returning to the chamber, each mode corresponds to a resonant frequency determined by the cavity dimensions. In the original treatments, each mode was considered to be a "harmonic oscillator" having, as we shall see, an average thermal energy kT. Before we start counting.the number of these modes *versus* optical frequency, let us first verify this *average* energy of a *single* mode according to Boltzmann's formula. First of all, we know that an ensemble of identical modes, either in time or over many systems, must have a total probability distribution over all energies U, which adds to unity, i.e.,

$$\int_0^\infty p(U)dU = \int_0^\infty Ae^{-U/kT}dU = 1 \quad \therefore A = \frac{1}{\int_0^\infty e^{-U/kT}dU} \tag{1.2}$$

The average energy of the mode is the integral over the product of the energy and the probability of that energy and is

$$\overline{U} = \int_0^\infty UAe^{-U/kT}dU = \frac{\int_0^\infty Ue^{-U/kT}dU}{\int_0^\infty e^{-U/kT}dU} = \frac{(kT)^2\int_0^\infty xe^{-x}dx}{kT\int_0^\infty e^{-x}dx} = kT \tag{1.3}$$

where we have used the mathematical relationship,

$$\int_0^\infty x^n e^{-x}dx = n! \tag{1.4}$$

We have thus obtained the standard classical result, which says that the energy per mode or degree of freedom for a system in thermal equilibrium has an average value of kT, the thermal energy. Soon we shall find that the number of allowed electromagnetic modes of a rectangular enclosure, or any enclosure for that matter, increases indefinitely with frequency. If each of these modes had energy kT, then the total energy would increase to infinity as the frequency approached infinity or the wavelength went to zero. This "ultraviolet catastrophe" as it was called, led to the proposal by Planck that at frequency, ν, a mode was only allowed discrete energies separated by the energy increment, $\Delta U = h\nu$. The value of the quantity, h, Planck's constant, was determined by fitting this modified theory to experimental measurements of thermal radiation.

Figure 1.1 shows the difference between the classical continuous Boltzmann distribution, (a), and a discrete or "quantized" distribution, (b). In the continuous distribution the *area* under the probability curve $p(U)$ is equal to unity. In the discrete or quantized case the allowed energies as shown by the bars are separated by $\Delta U = h\nu$ and the *sum of the heights of all bars* becomes unity. We may state this mathematically by writing the energy of the nth state as

$$U_n = nh\nu \qquad n = 0,1,2, etc.$$

with

$$p(U_n) = Ae^{-U_n/kT} = Ae^{-nh\nu/kT} \quad \therefore \sum_{n=0}^{\infty} Ae^{-nh\nu/kT} = 1 \tag{1.5}$$

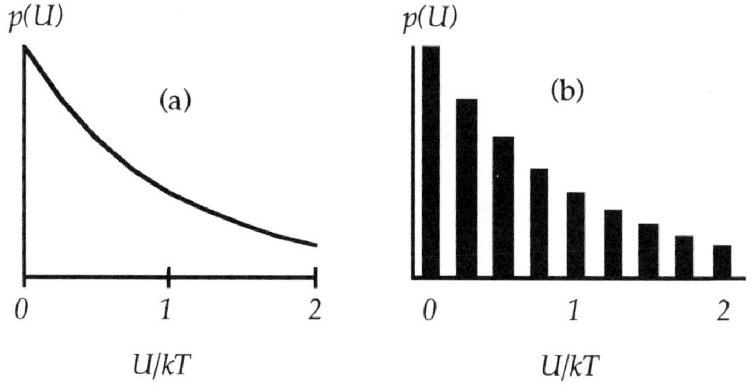

Figure 1.1 (a) Continuous and (b) discrete Boltzmann distribution with $\Delta U = h\nu = kT/4$.

In a similar manner we may calculate the average energy, $U(\nu)$, by summing the products of the nth state energy and its probability of occupation. Then

$$U(\nu) = \frac{\sum_0^\infty nh\nu e^{-nh\nu/kT}}{\sum_0^\infty e^{-nh\nu/kT}} = \frac{h\nu\sum_0^\infty nx^n}{\sum_0^\infty x^n}; \qquad x = e^{-h\nu/kT} \tag{1.6}$$

and using the identities,

$$\sum_0^\infty x^n = \frac{1}{1-x}; \qquad \sum_0^\infty nx^n = x\frac{d}{dx}\sum_0^\infty x^n = \frac{x}{(1-x)^2} \tag{1.7}$$

we finally obtain:

$$U(\nu) = h\nu\frac{x}{(1-x)} = h\nu\frac{e^{-h\nu/kT}}{(1-e^{-h\nu/kT})} = \frac{h\nu}{(e^{h\nu/kT}-1)} \tag{1.8}$$

This average energy for an electromagnetic mode at a single specific frequency, ν, now has a markedly different behavior from the classical result of Eq. (1.3) when the energy $h\nu$ becomes comparable to or greater than the thermal energy kT. In the two frequency limits, Eq. (1.8) goes to kT for low frequencies while it becomes $h\nu e^{-h\nu/kT}$ as the frequency becomes very large. Of major significance is that the ratio of $h\nu$ to kT for visible radiation at room temperature is of the order of one hundred, as we will see when we discuss the values of the various constants. As a result, the average energy per mode at visible frequencies is much less than kT.

The behavior of $U(\nu)$ can be understood by examination of Figure 1.1. As the spacing of the discrete energies becomes smaller and smaller, the distribution of energies approaches the classical form, while as the spacing increases, the probability of the mode being in the zero-energy state approaches unity, and the occupancy of the next state, $n = 1$ *or larger*, becomes negligibly small, and thus $U(\nu)$ goes to zero.

Given the expected energy for a single cavity mode at frequency, ν, we may calculate the energy density in an enclosed cavity by counting the number of available electromagnetic modes as a function of the frequency. We start with the rectangular chamber of Figure 1.2, of dimensions, a by b by d, which has walls at temperature, T. We then write the equation for the allowed electromagnetic standing wave modes subject to the condition that the electric field, E, goes to zero at the walls. This is

$$E = E_0 \sin(k_x x)\sin(k_y y)\sin(k_z z)\sin 2\pi\nu t \tag{1.9}$$

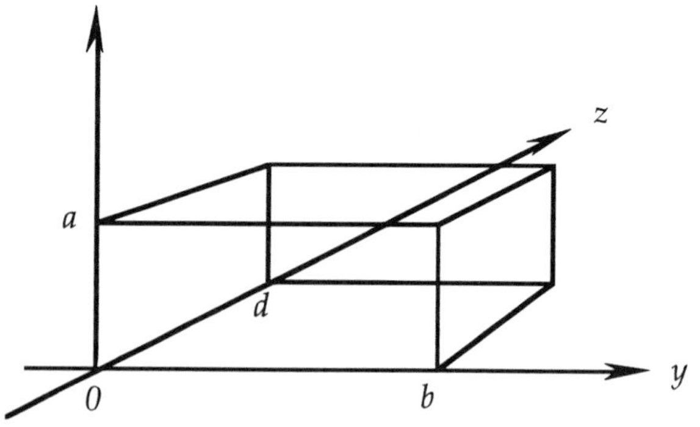

Figure 1.2 Rectangular box for calculation of mode densities.

with each k taking only positive values. Using Maxwell's wave equation,

$$\frac{\partial^2 E}{\partial x^2}+\frac{\partial^2 E}{\partial y^2}+\frac{\partial^2 E}{\partial z^2}=\frac{1}{c^2}\frac{\partial^2 E}{\partial t^2}$$

we obtain

$$k_x^2+k_y^2+k_z^2=\frac{4\pi^2\nu^2}{c^2}=\left(\frac{2\pi}{\lambda}\right)^2=k^2 \quad (1.10)$$

where c is the velocity of light and λ is the wavelength of the radiation. The quantity $k = 2\pi/\lambda$ is the the magnitude of the total wave vector for the particular mode. To determine the mode density *versus* frequency we use Figure 1.3, which is a representation of the allowed modes in k-space. These allowed modes occur at those values of k which cause the field to become zero at $x = a$, $y = b$, and $z = d$, since the sine function already produces a zero at $x = 0$, $y = 0$, and $z = 0$. The requisite values of k are respectively $m\pi/a$, $n\pi/b$, and $p\pi/d$, where m, n, and p are integers. The allowed modes thus form a rectangular lattice of points in k-space with spacing as shown in Figure 1.3. We now assume that the box dimensions are much greater than the wavelength λ and the distribution of points is then effectively continuous, since π/a for example is much less than $2\pi/\lambda$, the magnitude of the k-vector in Eq. (1.10).

We now determine the number of modes dN in a thin octant (or eighth of a sphere) shell of thickness dk by multiplying the density of modes by the volume of the shell. Since the radius of the shell is $k = 2\pi\nu/c$, all modes on its surface are at the same frequency ν. In addition, each point representing a mode lies on the corner of a rectangular volume with dimensions, π/a by π/b by π/d. Therefore the density of points is the inverse of this volume or $abd/\pi^3 = V/\pi^3$, where V is the volume of the box. The volume of the octant shell is one eighth of $4\pi k^2 dk$ so that

$$dN=\frac{V}{\pi^3}\cdot\frac{4\pi k^2 dk}{8}=\frac{4\pi V\nu^2 d\nu}{c^3} \quad (1.11)$$

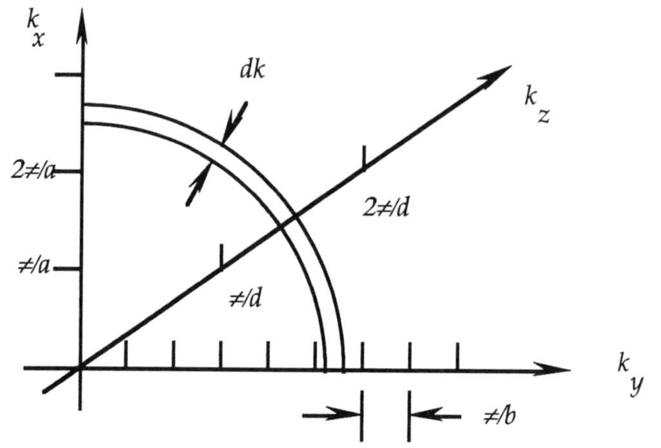

Figure 1.3 *k*-space showing discrete values of k_x, k_y, and k_z.

using the relation between k and ν from Eq. (1.10). Finally, we use the average energy per mode from Eq. (1.8) to calculate the energy density per unit frequency range, $u_\nu = du/d\nu$, where $u = U/V$, the electromagnetic energy per unit volume in a blackbody equilibrium cavity at temperature, T. In counting the modes we must take into account the two possible polarizations of the electric field, thus doubling the result of Eq. (1.11) and yielding

$$du = 2dN\frac{U(\nu)}{V} = \frac{8\pi\nu^2 d\nu}{c^3}\cdot\frac{h\nu}{(e^{h\nu/kT}-1)} = \frac{8\pi h\nu^3 d\nu}{c^3(e^{h\nu/kT}-1)}$$
$$= \frac{8\pi(kT)^4}{h^3c^3}\cdot\frac{x^3 dx}{(e^x-1)} \quad \text{with} \quad x = \frac{h\nu}{kT} \tag{1.12}$$

This is the fundamental Planck equation, which we have written in terms of the universal function $F(x) = x^3/(e^x\text{-}1)$ sketched in Figure 1.4. The energy density reaches a maximum at $x = h\nu/kT = 2.8$ and then the curve falls exponentially to zero.

Before we continue with our manipulations of Planck's law, we should discuss briefly the concept of the "photon," which is after all the heart of our topic. As we have reviewed, blackbody radiation was explained by Planck in terms of allowed discrete energies of an electromagnetic mode. In that context, a photon is a discrete step or quantum

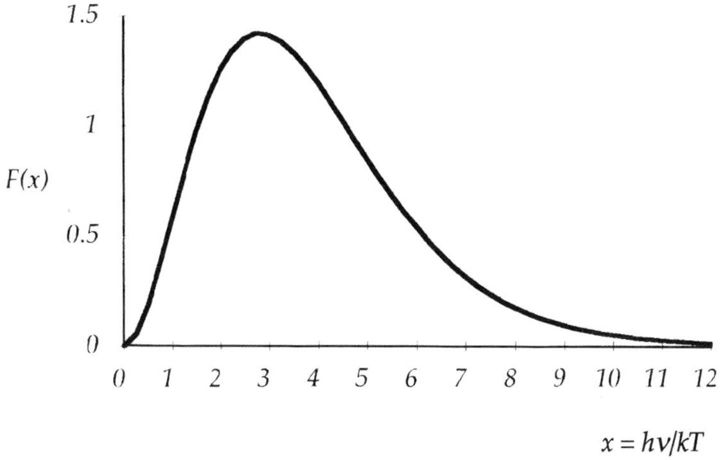

Figure 1.4 The Planck energy density function, $F(x)$.

of energy of magnitude $h\nu$. An alternative concept of the photon is that of a particle and the average energy of Eq. (1.8) may be written as the product of $h\nu$, the photon energy, and $1/(e^{h\nu/kT}-1)$, the occupation probability of the mode or the number of photons per mode. This probability factor is known in a more general form as the Bose-Einstein factor and is applicable in quantum mechanical treatments to "bosons" or particles with spin unity. Even though we soon speak of photon detectors, we shall use the term *photon* in the sense of a discrete energy gain or loss by the electromagnetic field, *never* as the description of a localized particle.

1.2 Energy Flow, Absorption, and Emission

We derived Planck's energy distribution with a rather idealized chamber with reflecting walls, which established the boundary condition on the electric field. We now generalize and state without proof that the energy density in *any* enclosed chamber obeys Eq. (1.12), even for partially or completely absorbing walls, provided that *all* interior surfaces are at the same temperature T. We also state that the energy density is

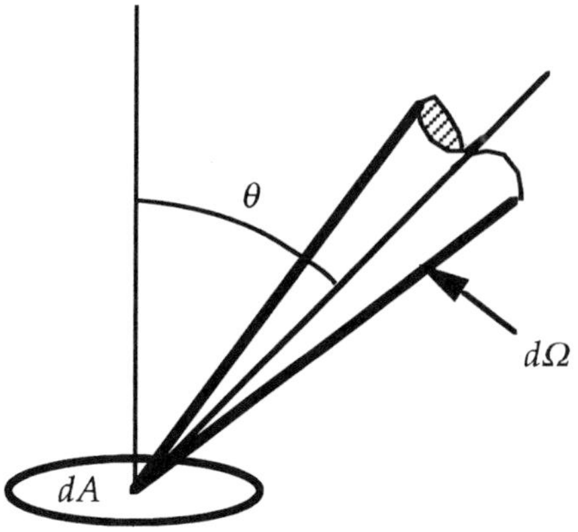

Figure 1.5 Power striking surface dA as a function of incidence angle θ.

uniform and *isotropic*, that is, the energy is flowing (at the velocity of light) in all directions with equal magnitude. These conclusions are a result of very general thermodynamic arguments and an excellent discussion may be found in *(Reif, 1965)*. We now find the incident power per unit area striking any surface of such a chamber and conclude with expressions for the absorbed as well as emitted power based on the surface properties and equilibrium temperature T. We first place an imaginary surface element dA at an arbitrary point in the cavity and determine the power crossing the surface in one direction, as shown in Figure 1.5. Now since the radiation is isotropic and moving at velocity c, the power density flowing toward the surface within solid angle $d\Omega$ is given by

$$dI_\nu = d\left(\frac{dI}{d\nu}\right) = cu_\nu \cdot \frac{d\Omega}{4\pi} \quad \text{with} \quad u_\nu = \frac{du}{d\nu} \tag{1.13}$$

where I_ν is the power per unit area per unit frequency, and u_ν is the energy per unit volume per unit frequency. [The *solid* angle Ω used frequently in our analysis, is measured in steradians and is given by A/R^2, where A is the area subtended by the angle on the surface of a sphere of radius R. Thus a full sphere contains 4π steradians, and as used in the following equation, Ω is equal to the area subtended by the solid angle on the surface of a *unit* radius sphere.] Power flowing toward the surface at an angle θ from the normal is intercepted by an effective area, $dA\cos\theta$, so that the power dP_ν striking *one* side of the surface from one hemisphere becomes

$$dP_\nu = \frac{dP}{d\nu} = \int_{\Omega} dA\cos\theta \cdot dI_\nu$$

resulting in

$$dP_\nu = \frac{cu_\nu dA}{4\pi} \cdot \int_0^{2\pi} \cos\theta \, d\Omega = \frac{cu_\nu dA}{4\pi} \cdot \int_0^{\pi/2} \cos\theta (2\pi \sin\theta \, d\theta) = \frac{cu_\nu dA}{4} \quad (1.14)$$

Finally, substituting for u_ν, we obtain the intensity or power per unit area per unit frequency received by the surface over 2π steradians of incidence angle. This is

$$I_\nu = \frac{dI}{d\nu} = \frac{dP_\nu}{dA} = \frac{cu_\nu}{4} = \frac{2\pi h\nu^3}{c^2(e^{h\nu/kT} - 1)} = \frac{2\pi (kT)^3}{h^2 c^2} \cdot F(x); \; x = \frac{h\nu}{kT} \quad (1.15)$$

using Eq. (1.12). This result gives us the *spectral irradiance* I_ν in *watts/m²Hz* striking *any* surface in the chamber such as the walls. Its frequency and temperature dependence is the same as that of the energy density distribution, with *F(x)* as shown in Figure 1.4.

Example: Let us calculate the *spectral irradiance* $I_\nu = dI/d\nu$ *versus* the frequency, in *W/m²Hz* for the two temperatures, $T = 300$ *and 600 K*. Substituting the values of *h*, *c*, and *k* (see Section 1.5) into the fourth term of Eq. (1.15) yields

$$\frac{dI}{d\nu} = \frac{2\pi h\nu^3}{c^2(e^{h\nu/kT} - 1)} = \frac{4.61 \times 10^{-50} \cdot \nu^3}{(e^{4.78\times10^{-11}[\nu/T]} - 1)}$$

which is plotted in Figure 1.6 for the two different temperature values. An important feature of these plots is the huge increase in integrated power over the frequency band, a factor of $2^4 = 16$ as we discuss in Section 1.3. Although a family of such curves can be used

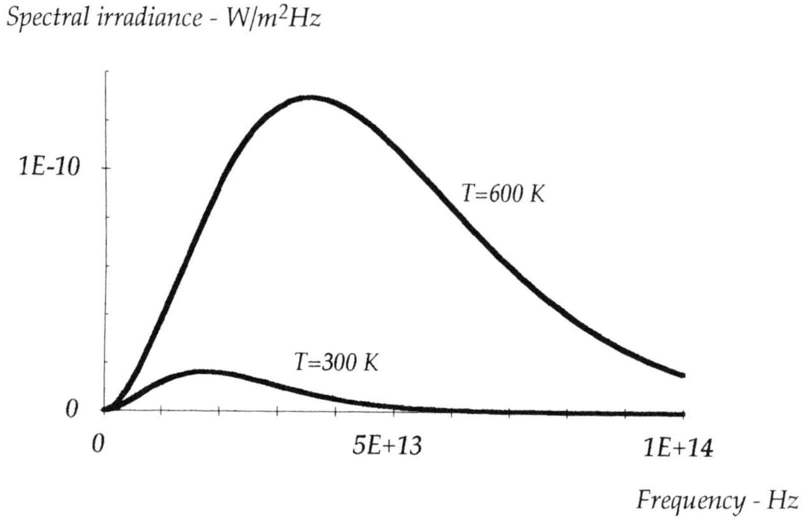

Figure 1.6 Spectral intensity *versus* frequency for $T = 300$ and $600\ K$. The frequency, $10^{14}\ Hz$ corresponds to a wavelength, $\lambda = 3\ \mu m$.

for calculating the energy within specified frequency bands, a more straightforward approach is given in Section 1.5.

We now assume that the wall of the chamber has a *reflectivity* ρ, whose value is a function of frequency ν and angle of incidence θ. The radiation is in equilibrium with the walls, so there is no net power flow, that is, the absorbed or nonreflected power is equal to the emitted power from the wall. The statistical law of detailed balance requires that this equality must hold at each frequency and angle of incidence. Since, by symmetry and the isotropic behavior of the radiation, the power leaving the surface must be equal to the power incident, then we must conclude that the reflected power plus the emitted power is equal to the incident power. If this is the case, then the emitted power is given by the quantity $(1 - \rho)$ times the incident power as given by equation (1.15). The quantity $(1 - \rho)$ is called the *emissivity*, and is expressed by the symbol ε, resulting in the equation

$$H_\nu = \varepsilon(\nu) I_\nu = \varepsilon(\nu) \frac{2\pi h \nu^3}{c^2 (e^{h\nu/kT} - 1)} \tag{1.16}$$

where H_ν, the *spectral radiance* of the surface over all frequencies, has the dimensions W/m^2Hz. The quantity H, the *radiance*, is the power per unit area radiated *from* the surface. We have made an assumption in this equation that the emissivity is independent of angle of emission, a reasonable one for most practical cases. The emissivity is always a function of optical frequency ν, and if equal to unity the surface is "black," or completely absorbing, leading to the term "blackbody." Thus a true blackbody at temperature T radiates power identical to the incident power within a fully enclosed chamber at temperature,T. Since the radiation coming from the surface is a property of the surface and its temperature, it follows that the *removal of the chamber has no effect on the radiance of the remaining surface element* as long as it is maintained at the specified temperature. This powerful conclusion allows us to calculate the emitted power from any object if we know its temperature and emissivity.

In a similar manner we may obtain the *radiance* of a partially absorbing medium such as the atmosphere. Here, as shown in Figure 1.7, we consider an absorbing layer of material with net *fractional power* transmission $T(\nu)$. If we now considered this layer as the wall between two blackbody enclosures, then we can conclude that the layer emits as a blackbody source as in Eq. (1.16) with $\varepsilon(\nu) = [1 - T(\nu)]$. This follows since the radiation transmitted through the layer from one enclosure to the other is unit emissivity blackbody radiation. Removal of the upper right chamber leaves only that radiation which replaced the *absorbed* incident radiation, or the fraction, $[1 - T(\nu)]$. We shall be concerned with the angular and spectral distribution of the radiance in later examples and problems but now consider the *total* radiation emitted over the whole frequency spectrum.

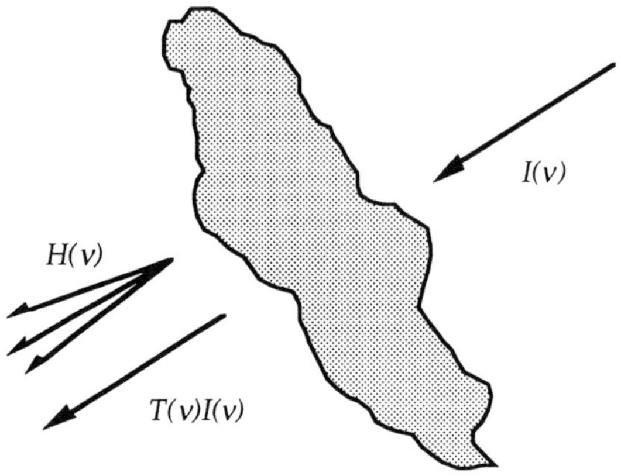

Figure 1.7. Model for calculation of background radiation from a partially transmitting medium.

1.3 The Stefan-Boltzmann Law

We now calculate the *total* radiance H of a surface of *constant* emissivity ε at temperature T; the result is known as the Stefan-Boltzmann law. Starting with Eq (1.16), we obtain

$$H = \int H_\nu d\nu = \varepsilon \int_0^\infty \frac{2\pi h\nu^3 d\nu}{c^2(e^{h\nu/kT} - 1)} = \varepsilon \frac{2\pi(kT)^4}{h^3 c^2} \int_0^\infty F(x)dx; \quad F(x) = \frac{x^3}{(e^x - 1)} \tag{1.17}$$

Fortunately, the dimensionless integral has an explicit value of $\pi^4/15$, and the final result may be written

$$H = \varepsilon \frac{2\pi^5 k^4}{15c^2 h^3} T^4 = \varepsilon\sigma T^4 \tag{1.18}$$

and the total irradiance striking the walls of a blackbody chamber is $I = \sigma T^4$, where σ, the Stefan-Boltzman constant, is

$$\sigma = 5.67 \times 10^{-8}\ W/m^2\ K^4$$

For a unit emissivity surface at *300 K*, the radiance becomes *460 W/m²*. In comparison with this value for room-temperature radiation we may calculate the radiance of the sun's surface, at *5800 K*, to be 6.42×10^7 *watts/m²*. The rapid fourth-power increase in radiance with temperature is a result of the combination of a linear increase in the frequency maximum with temperature combined with a cubic increase in total mode energy, since the mode density goes as the square and the mode energy directly with frequency.

Next we wish to calculate another more pertinent number with regard to the sun. This is called the *solar constant,* which is the total intensity in *W/m²* of the solar radiation striking the earth above any absorbing atmosphere. To obtain the value of this quantity, we return to our blackbody chamber and assume the walls have unit emissivity and are heated to *5800 K*. In this case, however, as shown in Figure 1.8, we assume a hemispheric projection from the upper wall which represents the sun. Just as we removed the whole chamber to determine the power emitted from a surface, we shall now remove all but the hemisphere and determine the irradiance *I*, which is striking a point at the bottom of the chamber. Since Ω for the sun is small, all the radiation is effectively normal to the surface and, using Eq. (1.14), the irradiance becomes

$$(I_\nu)_{solar} = \frac{dP_\nu}{dA} = \frac{cu_\nu \Omega}{4\pi} = \frac{\Omega}{\pi} I_\nu \qquad (1.19)$$

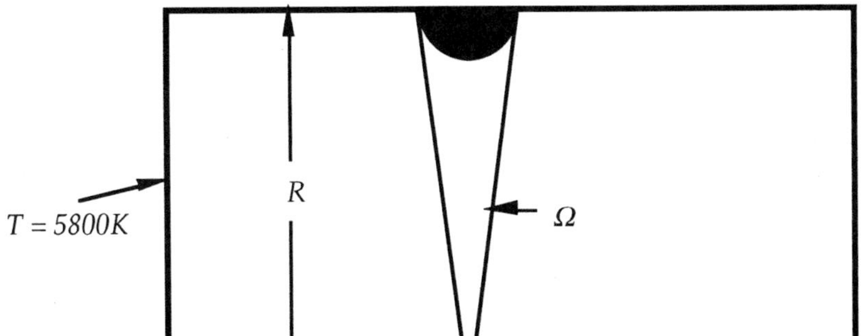

Figure 1.8 Calculation of solar constant

from Eq. (1.14). The irradiance integrated over all frequencies is then

$$I = \frac{\Omega}{\pi} \cdot \sigma T^4 \tag{1.20}$$

Our derivation is based on the unit emissivity of this "sun." This guarantees that all the power striking the bottom of the chamber is emitted from the "sun" rather than reflected off of it from the adjacent walls, which we removed for our calculation. It is also important to note that the actual *shape* of the surface is not critical as long as it subtends the same solid angle Ω. We discuss this matter further when we calculate the image plane intensity for an optical system. Returning to Eq. (1.20) we now calculate the solar constant, using the angular diameter of the sun, *9.3 milliradians*, and the radiance, $\sigma T^4 = 6.42 \times 10^7\ W/m^2$. The final result is

$$I = \frac{\Omega}{\pi} \sigma T^4 = \frac{(\pi/4)(9.3 \times 10^{-3})^2}{\pi} \cdot 6.42 \times 10^7 = 1390 W/m^2$$

where we have used the small-angle approximation, $\Omega = (\pi/4)\,\theta^2$, with θ the solar angular *diameter*. This number is comparable to the value of the radiance from a *300 K* blackbody. The significance is treated in one of the problems.

1.4 Image Intensity in an Optical Receiver

Throughout this text we shall be concerned with the collection of radiation by an optical system and its detection in the image plane of the receiver. Many of the sources we consider are either blackbody, sometimes called "graybody" if ε is constant but less than one, or they are ideal diffuse scatterers, which reflect incident light uniformly in all directions. Consider the *small* radiating area element A in Figure 1.9. The resultant intensity or irradiance I_S at the surface S will be proportional to the projected area as seen at the irradiated surface which is $A\cos\theta$. This is sometimes easier to understand if we think of the area element as a square hole in the surface of a blackbody chamber. Now if we consider the full hemispherical surface, S, of radius, R, en-

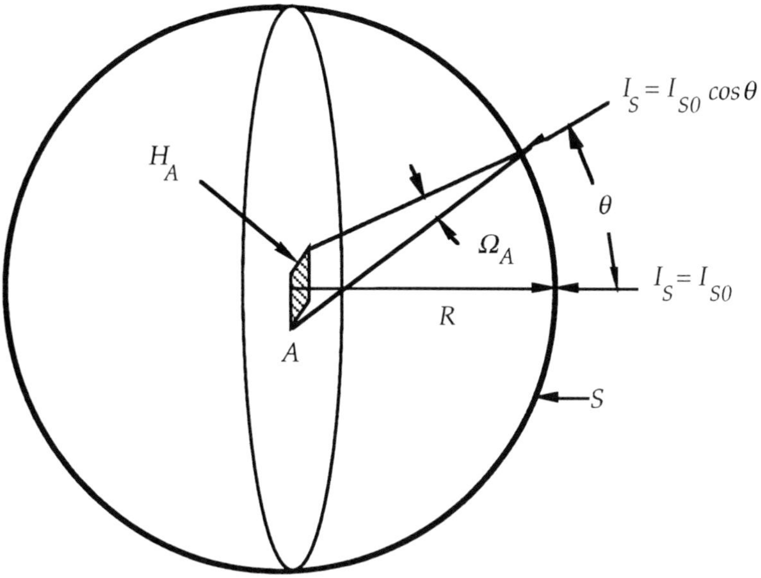

Figure 1.9 Intensity at range, *R*, for blackbody or diffuse source

closing *A*, the power radiated by *A*, $P = H_A A$, must equal the total power crossing the hemispherical surface at range *R*. Therefore,

$$P = H_A A = \int_0^{2\pi} I_{S0}\cos\theta R^2 d\Omega_S = \int_0^{\pi/2} I_{S0}\cos\theta\, R^2\, 2\pi\sin\theta\, d\theta = \pi I_{S0} R^2 \tag{1.21}$$

where Ω_S is the solid angle subtended by *S* about the small area element *A*. The intensity at the surface *S* then becomes

$$I_S = I_{S0}\cos\theta = \frac{H_A A\cos\theta}{\pi R^2} = H_A\frac{\Omega_A}{\pi} \tag{1.22}$$

where Ω_A is the solid angle subtended at *S* by the small area element as shown in Figure 1.9. Note that this equation is equivalent to Eq. (1.20) for the solar constant. Alternatively, element *A* could be a "perfectly diffuse" reflector with reflectance, ρ. In this case the radiance of the source becomes $H_A = \rho I_n$ where I_n is the *normal* component of the incident radiation, since it is this component that determines the total power incident on the surface. Again, Eq. (1.22) applies since the scattered radiation is isotropic or uniform in all directions just as in ideal blackbody radiation. Such a diffuse scatterer is often called a "Lambertian" scatterer since it obeys Lambert's law, which states that a surface with isotropic emissivity or a diffuse surface with uniform incident radiation appears to have the same "brightness" independent of the viewing angle. Brightness is defined as the observed power per unit area per unit solid angle at the receiver and determines the incident intensity at the focal plane of an optical system or at the retina of the eye.

Example: How bright is moonlight ? The irradiance at the earth's surface produced by a full moon may be calculated using Eq. (1.22) as follows:

Approximating the moon by a flat diffuse reflector (actually the moon is *not* a true diffuse reflector nor is it flat), we may obtain the moon's radiance H as the reflectance, ρ, multiplied by the solar constant, since the moon is effectively at the same distance from the sun as is the earth. Then from Eq. (1.22), we may calculate the "full moon lunar constant" as

$$I = \Omega H / \pi = [(\pi\theta^2/4)/\pi]\rho I_{solar}$$

where θ is the lunar angular diameter. Using an angular diameter of *9.3 milliradians*, an estimated reflectance of *0.5*, and a solar constant of *1390* W/m^2 yields an irradiance 1.5×10^{-2} W/m^2, or very close to 10^{-5} solar constants. This number gives interesting information on the dynamic range of the human eye. (The reader is asked to find how we obtained the lunar angular diameter.)

We now consider the simple optical system of Figure 1.10, which consists of a Lambertian source of radiant power P_S and area A_S, producing an image in the focal plane of area A_i with power P_i. The lens diameter is d and with a very large range R, the image is formed at the focal length of the lens f. Any image point may be determined by drawing a straight line from the desired object point through the *center* of the lens terminating at a distance, f, to the right of the lens. This rule

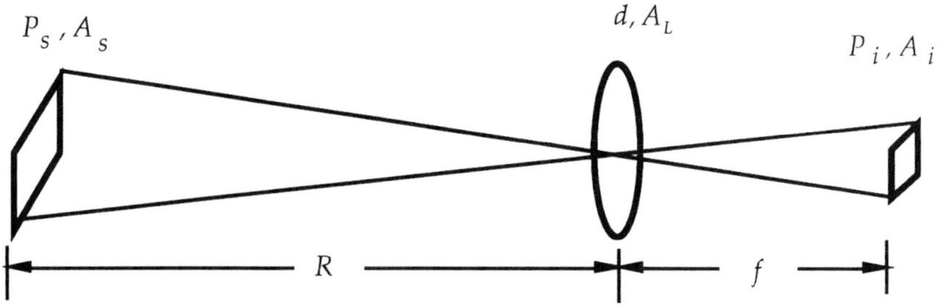

Figure 1.10 Simple optical system.

results from the non-deflection of any ray at the lens center since the front and rear surfaces are parallel at this point. For small angular deviations, the position in the focal plane is the product of the angle of incidence in *radians* and the focal length. The total power incident on the lens is

$$P_L = \frac{P_s}{A_s}\frac{\Omega}{\pi} A_L = \frac{P_s}{A_s}\left(\frac{A_s}{\pi R^2}\right) A_L = P_s \frac{A_L}{\pi R^2} \tag{1.23}$$

and this power is deposited in an image of area $A_i = (f/R)^2 A_S$ with $P_i = P_L$. With $A_L = \pi d^2/4$, we obtain

$$I_i = \frac{P_i}{A_i} = \frac{P_s}{A_s} \times \frac{A_L}{\pi f^2} = H_s \frac{d^2}{4f^2} = H_s \frac{1}{(2f/\#)^2} \tag{1.24}$$

where $f/\#$ is the *f*-number or *f*-stop familiar to the photographer. For a simple lens with f much greater than d, $f/\# = f/d$. For a "fast" lens or general optical system where $f/\#$ approaches unity, $f/\# = 1/(2\sin\theta)$, where $\sin\theta$, known as the *numerical aperture* (*NA*) is one-half the convergence angle of rays striking the image surface. Thus the maximum value of $\sin\theta$ is unity, corresponding to an $f/\#$ of 0.5, and image illumination from a full hemisphere having a convergence angle of 180°. Figure 1.11(a) illustrates the definition of the numerical aperture and Fig. 1.11(b) shows a possible $NA = 1$, $f/\# = 0.5$, system using a parabolic mirror. In either example, a small angular displacement of the incident

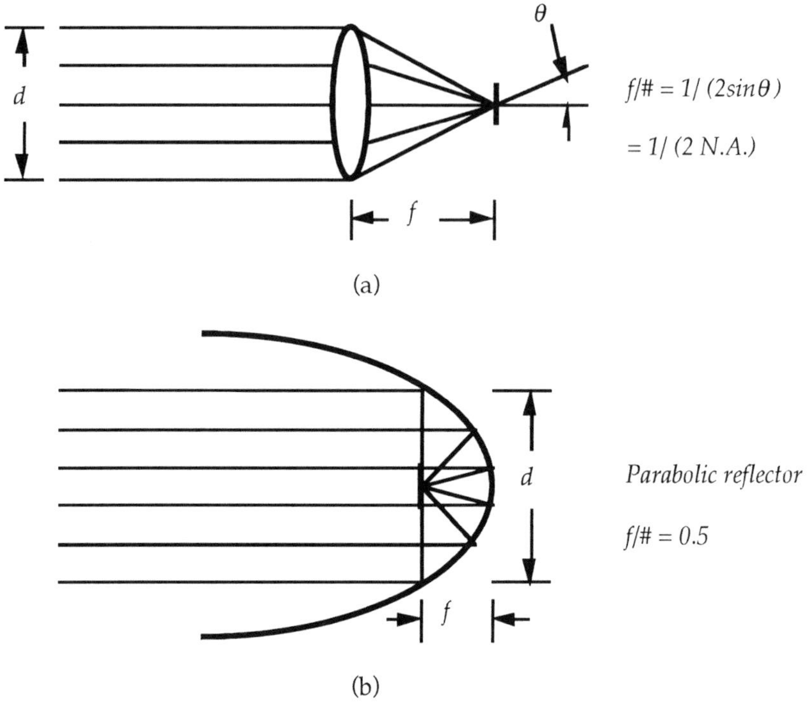

Figure 1.11 (a) Determination of *f/#* and *NA*. (b) Example of *f/#* = *0.5* system using a parabolic mirror.

rays, α, results in a change in focal plane position of $f\alpha$. For large f/d ratios, $\sin\theta$ becomes $d/2f$ and the *f/#* thus f/d. Although our derivation of the image intensity assumed large f/d ratios, the *last* term in Eq. (1.24) is still the correct expression for any *f/#* system. An important conclusion from Eq. (1.24) is that the intensity of the image is always less than the intensity of the source and that for an optical system that deposits energy on only one side of the focal plane, the allowable minimum *f/#* of *0.5* corresponds to equal source and image intensities. Were this not the case, one could cause the intensity of the focal spot to rise above the temperature of the source, a violation of thermodynamic laws.

1.5 Units and Intensity Calculations

We close this chapter by first listing the important numerical constants that are necessary for our calculations. These are

q	electron charge	1.6×10^{-19} *coulombs*
k	Boltzmann's constant	1.38×10^{-23} *joules/K*
h	Planck's constant	6.6×10^{-34} *joule-seconds*
c	velocity of light	3.00×10^{8} *meter/second*

We also use wavelength and frequency interchangeably and note that the wavelength λ is given by c/ν. Figure 1.12 encompasses the frequencies and wavelengths that are of concern to us, namely that region from the infrared to the ultraviolet where there is usable atmospheric transmission and where the photon energy $h\nu$ is much greater than the thermal energy kT at room temperature.

For convenience in calculations it is often helpful to measure these energies in electron volts, eV, the energy gained by an electron through a potential drop of one volt, 1.6×10^{-19} J. In this measure there are two simple expressions to remember. These are

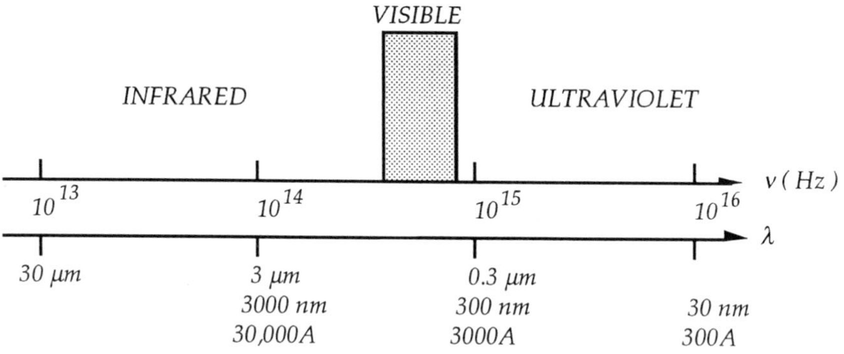

Figure 1.12 Infrared, visible, and ultraviolet frequencies and wavelengths.

$$h\nu\,(eV) = h\nu/q = hc/q\lambda = 1.24/\lambda\,(\mu m)$$

$$kT(eV) = kT/q = 0.026\;(T/300)$$

Thus, at room temperature, $300\ K$, the thermal energy kT is $0.026\ eV$, while the "photon" energy at a wavelength of $1\ \mu m$ or $10{,}000$ *angstrom units*, is 48 times as large or $1.24\ eV$. This is why there is effectively zero visible light radiated by room temperature objects.

A final help in performing radiation calculations is an expression for the fractional amount of radiated power from a blackbody occurring at wavelengths shorter than a given value λ. This may be obtained by evaluating the expression

$$I(>x) = \frac{15}{\pi^4}\int_x^{\infty}\frac{y^3dy}{(e^y-1)}; \quad x = \frac{h\nu}{kT} \tag{1.25}$$

obtained from Eq. (1.17). Here $I(>x)$ has a maximum value of unity and decreases as the minimum photon energy increases from zero. Since, as we will learn, most detectors only respond above a given photon energy (or below a given "cut-off" wavelength), the appropriate spectrum-limited radiance may be obtained by taking the product, $I(>x)\varepsilon\sigma T^4$. The emissivity ε should of course be an appropriate weighted average. Since $F(x)$ falls off almost exponentially at high photon energies, a choice of the emissivity at the cut-off wavelength usually gives an accurate answer. The function $I(>x)$ is plotted in Fig. 1.13.

Example: Calculate the power in W/m^2 radiated by a unit emissivity surface at temperature $2000\ K$ at wavelengths less than $1\ \mu m$. First,

$$\sigma T^4 = 5.67 \times 10^{-8}(2000)^4 = 9.1 \times 10^5\ W/m^2.$$

and we then find $x = h\nu/kT$ from $h\nu = (1.24/1) = 1.24\ eV$ and $kT = 0.026(2000/300) = 0.17\ eV$ The argument, x, becomes 7.3 and $I(>x) = 0.067$. The final radiance is then 6×10^4 or $60\ kW/m^2$.

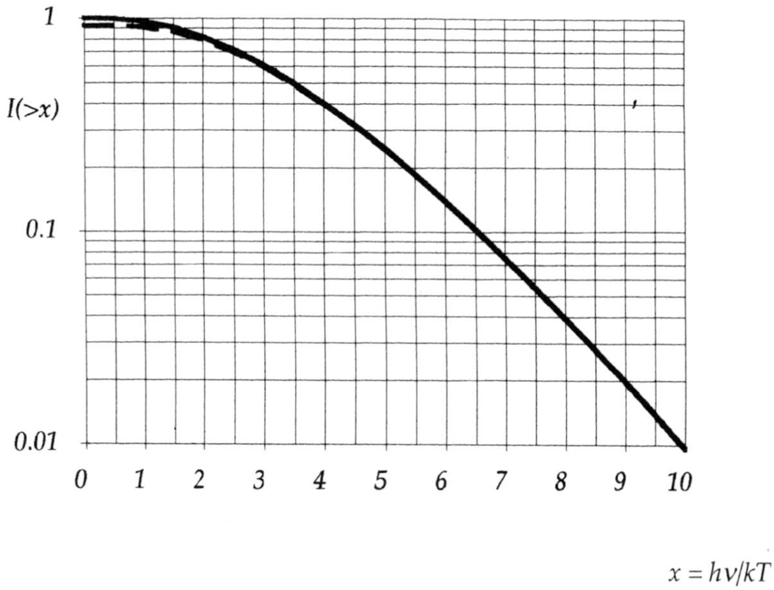

Figure 1.13 Fractional radiated power above a photon energy, xkT. The dashed line is the approximation of Eq. (1.26).

An excellent approximation for $I(>x)$ is (see Problem 1.13)

$$I(>x) = \frac{15}{\pi^4} x^3 e^{-x} \left[1 + \frac{3}{x} + \frac{6}{x^2} + \frac{6}{x^3} \right] \tag{1.26}$$

Shown in Figure 1.13, this approximation is within 6% for $x \geq 1$ and is within 1% for $x > 4$.

An alternative way to calculate the radiance and irradiance is to determine the number of *photons per second per meter*^{2}. In this case the *photon radiance* H^ϕ becomes from Eq. (1.17),

$$H^\phi = \int \frac{H_\nu}{h\nu} d\nu = \varepsilon \int_0^\infty \frac{2\pi\nu^2 d\nu}{c^2(e^{h\nu/kT} - 1)} = \varepsilon \frac{2\pi(kT)^3}{h^3c^2} \int_0^\infty G(x)dx; \; G(x) = \frac{x^2}{(e^x - 1)} \tag{1.27}$$

and the dimensionless integral becomes, using Eq. (1.4),

$$\int_0^\infty \frac{x^2}{e^x - 1}dx = \sum_{n=1}^{\infty}\int_0^\infty x^2 e^{-nx}dx = 2\sum_{n=1}^{\infty}\frac{1}{n^3} = 2.404 \quad (1.28)$$

and the photon radiance over the full spectrum becomes

$$H^\phi = \varepsilon \frac{2\pi(kT)^3}{h^3c^2}(2.404) = \varepsilon\left[\frac{\sigma\ (2.404)}{k\,(\pi^4/15)}\right]T^3 = \varepsilon(1.521\times 10^{15})T^3 \quad (1.29)$$

assuming a constant emissivity ε. Thus the photon radiance of unit emissivity source at $300\ K$ becomes 4.1×10^{22} *photons/m²sec*. We may also calculate the net photon radiance below a given photon energy by writing, similar to Eq. (1.25),

$$J(>x) = \frac{1}{2.404}\int_x^\infty \frac{y^2dy}{e^y - 1}; \quad x = \frac{h\nu}{kT} \quad (1.30)$$

which is plotted in Figure 1.14. The photon radiance then becomes

$$H^\phi = J(>x)\varepsilon(1.521\times 10^{15})T^3 \quad (1.31)$$

and the approximation,

$$J(>x) = \frac{1}{2.404}x^2e^{-x}\left(1 + \frac{2}{x} + \frac{2}{x^2}\right) \quad (1.32)$$

is shown by the dashed line in Figure 1.14.

Example: Consider again a unit emissivity source at $2000\ K$. The photon radiance at wavelengths shorter than $1\ \mu m$ becomes

$$H^\phi = J(>x)\varepsilon(1.521\times 10^{15})T^3 = (0.02)(1.521\times 10^{15})(2000)^3$$
$$= 2.43\times 10^{23}\ \ photons/m^2\ sec$$

Since most of the radiation lies very close to the cut-off wavelength,

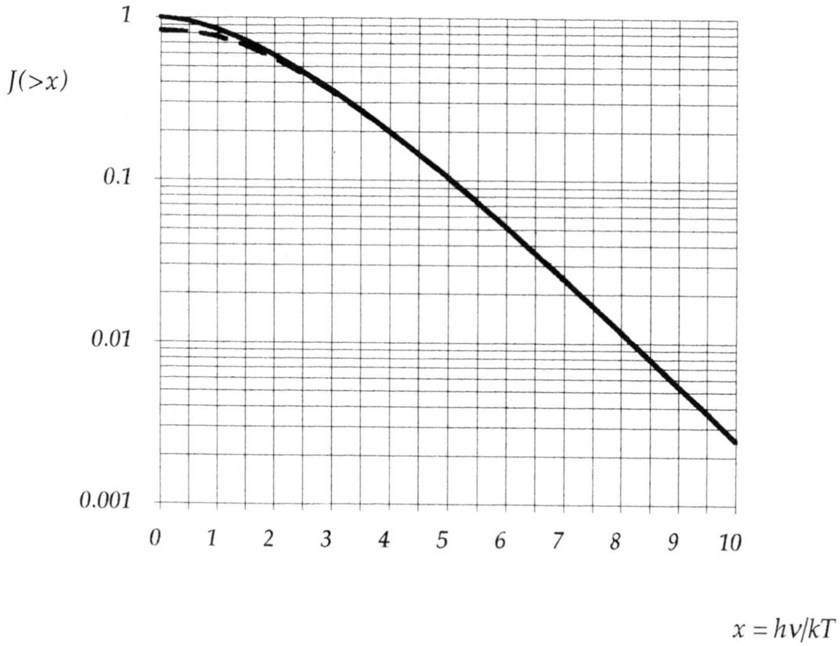

Figure 1.14 Fractional photon radiance above a photon energy, xkT. The dashed line is the approximation of Eq. (1.32).

a good approximation is to divide the radiance from the previous example, $6 \times 10^4\ W/m^2$ by the photon energy at $1\ \mu m$: $(1.24)(1.6 \times 10^{-19}) = 2.0 \times 10^{-19}$ *joules*. The result is 3.0×10^{23} *photons/m²sec*, very close to the exact value, due to the exponential decrease of radiance with increasing photon energy.

Another useful expression is the *spectral radiance* of a source, $H_\nu = dH/d\nu$, or in terms of wavelength λ, $H_\lambda = dH/d\lambda$. We write, from Eqs. (1.17) and (1.18),

$$dH = \varepsilon\sigma T^4 \frac{15}{\pi^4} F(x)dx = \varepsilon\sigma T^4 \frac{15}{\pi^4} F(x)\left[x\frac{d\nu}{\nu}\right] = \varepsilon\sigma T^4 \frac{15}{\pi^4} F(x)\left[x\frac{d\lambda}{\lambda}\right] \tag{1.33}$$

yielding for the two forms of *spectral radiance*

$$H_\nu = \frac{dH}{d\nu} = \varepsilon\sigma T^4 \frac{15}{\pi^4} F(x)\frac{x}{\nu} = \frac{\varepsilon\sigma T^4}{\nu}\frac{15}{\pi^4}\frac{x^4}{(e^x-1)} = \frac{\varepsilon\sigma T^4}{\nu}K(x)$$
$$H_\lambda = \frac{dH}{d\lambda} = \varepsilon\sigma T^4 \frac{15}{\pi^4} F(x)\frac{x}{\lambda} = \frac{\varepsilon\sigma T^4}{\lambda}\frac{15}{\pi^4}\frac{x^4}{(e^x-1)} = \frac{\varepsilon\sigma T^4}{\lambda}K(x) \tag{1.34}$$

The function,

$$K(x) = \frac{15}{\pi^4}\frac{x^4}{(e^x-1)} \tag{1.35}$$

is plotted in Figure 1.15. In a similar fashion, the spectral radiance of a diffuse surface illuminated *by* a blackbody source becomes

$$H_\nu = \frac{dH}{d\nu} = \frac{H}{\nu}\frac{15}{\pi^4}\frac{x^4}{(e^x-1)} = \frac{\rho I_n}{\nu}\frac{15}{\pi^4}\frac{x^4}{(e^x-1)} = \frac{\rho I_n}{\nu}K(x)$$
$$H_\lambda = \frac{dH}{d\lambda} = \frac{H}{\lambda}\frac{15}{\pi^4}\frac{x^4}{(e^x-1)} = \frac{\rho I_n}{\lambda}\frac{15}{\pi^4}\frac{x^4}{(e^x-1)} = \frac{\rho I_n}{\lambda}K(x) \tag{1.36}$$

where ρ is the surface reflectivity and I_n is the *normal* component of the incident irradiance.

Example: A white surface is illuminated by the sun with an incidence angle from the normal of 60°. Find the spectral radiance in $W/m^2\ \mu m$ at a wavelength of $\lambda = 1.06\ \mu m$. For this case,

$$x = h\nu / kT = [1.24/1.06] \div [0.026(5800/300)] = 1.19/0.502 = 2.37$$

The spectral radiance then becomes

$$H_\lambda = \frac{dH}{d\lambda} = \frac{H}{\lambda}K(2.37) = \frac{\cos 60^o(1390)}{1.06}(0.50) = 328\frac{W}{m^2\ \mu m}$$

and we note that the wavelength unit for the spectral density (μm^{-1}) matches the unit used for λ (μm). The value obtained appears at first

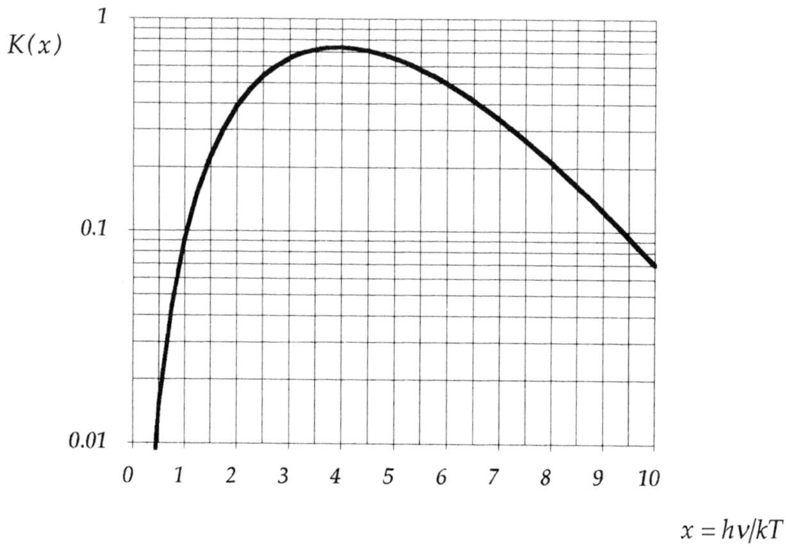

Figure 1.15 Function, $K(x)$, for calculation of spectral radiance

to be quite large until one realizes that a spectral *width* of $1\ \mu m$ is of the order of 100% of the wavelength. A typical narrowband spectral filter might have a bandwidth of $50\ A$ or $0.005\ \mu m$, yielding an effective radiance of $1.64\ W/m^2$.

Although we seldom use them in this text, an alternate set of units exists, called *photometric*. These units describe the intensity of optical radiation in terms of the response of the human eye *(Boyd, 1983, Ch. 6)*. *Power* becomes *luminous flux* and is measured in *lumens*. *Irradiance* becomes *illuminance* and is measured in *lumens*/m^2 or *lux*. We do not review the many other *photometric* units, but here describe the relation between *watts* and *lumens*. Since these units are a measure of the apparent optical power as observed by the eye, they are based on an empirical international standard curve of eye sensitivity called the luminosity factor, $Y(\lambda)$, shown in Figure 1.16. The conversion from power $P(\lambda)$, in watts, to luminous flux $L(\lambda)$, in lumens, is given by

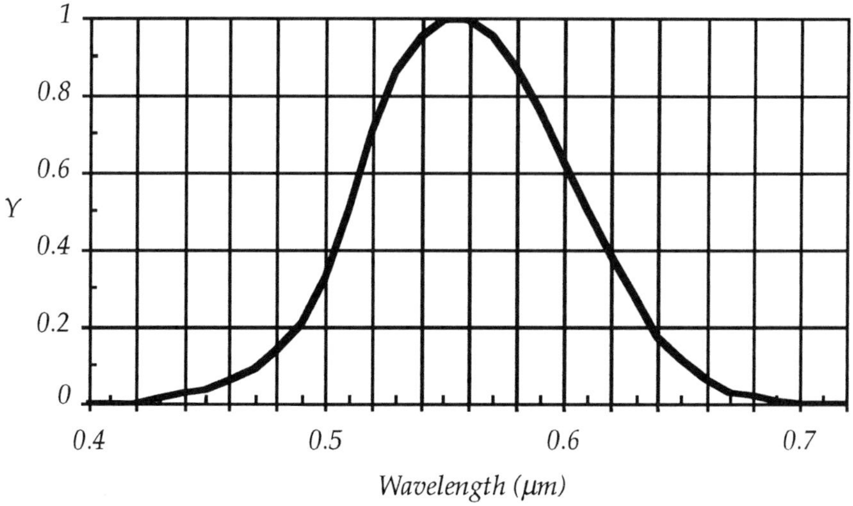

Figure 1.16 The relative luminosity factor, $Y(\lambda)$.

$$L(\lambda) = 680\int_0^\infty P(\lambda)Y(\lambda)d\lambda \tag{1.37}$$

where the coefficient, *680*, is part of the empirical definition of the *lumen*. The luminous efficiency, in *lumens/watt*, is thus *680*, if *all* the radiation is at *0.55 μm*, and is always less for a source with a distributed spectrum. As examples, consider first solar illumination, where the solar constant is $1400\ W/m^2$, which is tabulated *(Boyd, 1983)* as 1.2×10^5 $lumens/m^2$ or *lux*. Here the luminous efficiency is *85 lumen/W*, or a relative efficiency of *85/680* or *12.5%*. In contrast, a *60-watt* incandescent light bulb produces approximately *1000 lumens*, giving a luminous efficiency of *17 lumens/W* or a relative efficiency of *2.5%*. The reason for this reduced efficiency is, of course, the lower temperature of the source, *2800 K* for the bulb *versus 5800 K* for the sun, and thus the shift of the light bulb spectrum further away from the peak response of the eye.

Problems

1.1. Some visible light detection systems are limited by background or extraneous light from the surroundings. Usually *300K* blackbody radiation is not a problem. To verify this:

(a) Calculate what fraction of the power radiated at *300 K* occurs at wavelengths shorter than *1* μm. Use the approximation of Eq. (1.26).

(b) For wavelengths shorter than *1* μm, how many *photons/second* are emitted by a *1-*m^2 source with unit emissivity at *300 K*. Use a photon energy corresponding to a wavelength of *1* μm.

1.2 A long two-wire (single polarization mode) transmission line of length, *L*, acts as a resonator with modes determined by $n\lambda/2 = L$, with *n* an integer. Find the mode density in the frequency domain and then find the thermal energy per unit length per unit frequency in terms of kT and $h\nu$. What is the limiting form for low-frequency, that is, $h\nu << kT$?

1.3 There is an atmospheric transmission band from *8* to *12* μm that is near the peak of the *300 K* blackbody spectrum. Estimate the total power radiated from a *1-*m^2 unit emissivity surface in both *watts* and *photons/second*, within this band. Approximate by calculating the power spectral density at *10* μm and multiplying by the bandwidth. Compare with a calculation based on Figure 1.13. Use the average energy of the photons at *10* μm.

1.4 The *reflectivity* of the earth averaged over the solar spectrum, sometimes called the earth albedo (literally, *whiteness*), is on the average *0.35*. Using the solar constant and an earth temperature of *300K*, find the effective emissivity ε averaged over a *300K* spectrum. Remember that the earth receives solar energy from one direction but radiates its thermal energy isotropically.

1.5 If the emissivity of the earth decreases by *10%* of its current value, find the rise in average temperature, assuming an unchanged albedo. You may approximate by using differential forms dP and dT.

1.6 The radii of the orbits of Mercury, Earth, and Pluto are respectively *36, 93,* and *3700 million miles*. With Earth at *300 K*, find the temperature of the other two planets assuming the solar reflectivity and average emissivity are the same as those of earth.

1.7 Two planes of unit emissivity are at temperatures T_1 and T_2. They are separated by a vacuum region as shown in the sketch. If $(T_1 - T_2)$ is small and both are near *300 K*, find a numerical value for the radiative heat conductance, $G_H = (1/A)dP/dT$, for heat flow between the planes. State your answer in W/m^2K.

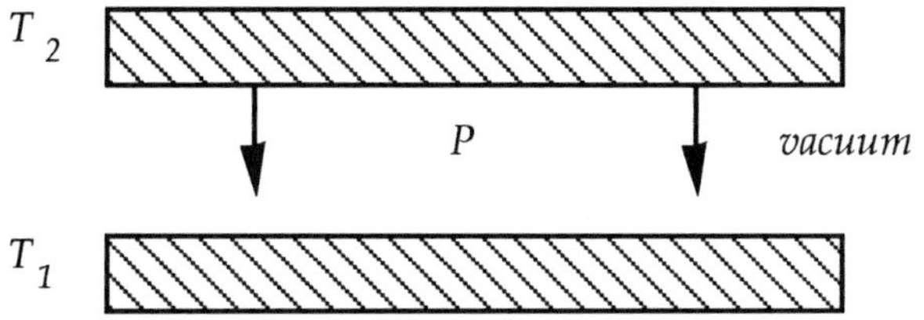

1.8 One of the legendary causes of fire is that due to glasses (spectacles) left on the window sill on a sunny day. Assume a *4 cm* lens diameter and a focal length of *1 m* (the optician would call this a one diopter , or $1/f = 1\ m^{-1}$ correction). Let the sun shine normal to the lens and strike a perfectly absorbing surface at the focal distance. If the surface can only lose heat by radiation from the incident side, find its temperature.

1.9 You are performing medical thermography (remote sensing of skin temperature) using an infrared camera operating in the wavelength band of *8* to *12* μm. Using the same approximation as Problem 1.3, calculate the fractional change in the emitted power with temperature, T. Specifically, find $(1/P)(dP/dT)$.

1.10 Continuing from Problem 1.9, your camera has a $10\text{-}cm^2$ lens of focal length *10 cm* and is observing the patient at a range of *1 m*. A detector in the image plane "sees" a skin area of $1\ cm^2$. Assume the skin has unit emissivity and calculate the number of photons received by the detector in a *1-msec* exposure time.

1.11 When we discuss photon detectors, we will find that for a perfect detector, the count fluctuation in an ensemble of measurements, the *rms* error, is the square root of the average count. If this is the case find that temperature change ΔT in problem 1.10 that will give a change in photon count equal to the fluctuation or uncertainty in the measurement. (This quantity is often called the "noise equivalent temperature" $NE\Delta T$.)

1.12 A camera set at *f/16* photographs a unit reflectivity diffuse surface, which is illuminated at an angle of *45 degrees* by the sun. Find the number of $photons/cm^2$ available to expose the film with an exposure time of *1 msec*. Assume *35%* of the solar energy is effective in film exposure with an average wavelength of *5000 A (0.5 μm)*.

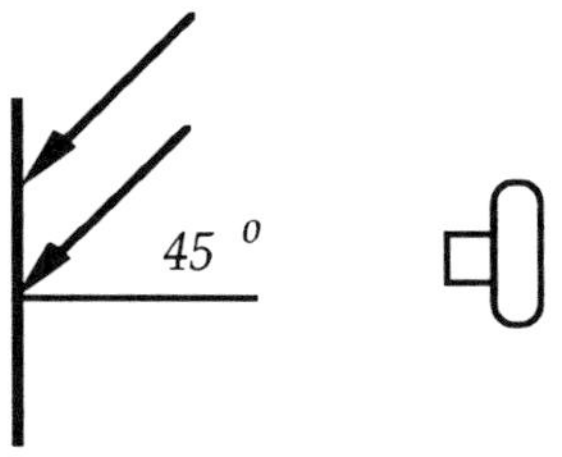

1.13 Show that for $e^x \gg 1$,

$$\int_x^{\infty} \frac{y^3 dy}{(e^y - 1)} = x^3 e^{-x} \cdot \left[1 + \frac{3}{x} + \frac{6}{x^2} + \frac{6}{x^3}\right]$$

Hint: Change the variable and use Eq. (1.4).

1.14 The eye's spectral sensitivity can be approximated by a square response curve from *5000* to *6250 A (0.5 to 0.625 μm)*. Calculate the fractional visible radiated power from a unit emissivity tungsten headlamp operating at *2800 K* and compare it with that of the newer tungsten-halogen headlamps which operate at *3200 K*.

1.15 A dark-adapted eye can see a heated glowing object if it is at a temperature T or higher. Estimate this temperature given the following information:

The retina or photosurface has a circular resolution element of *6 μm* diameter and requires *100 photons* at a wavelength $\lambda = 0.6\ \mu m$ or shorter for detection. When dark adapted, the effective $f/\# = 3$, and the integration time is *0.1 seconds*. *Suggestion:* Find the value of x in the approximation of Eq. (1.26) by successive approximations.

1.16 A He-Ne supermarket laser ($\lambda = 0.63\ \mu m$) scans a bar code at a distance of *30 cm* from a $1\ cm^2$ collecting lens collocated with the laser. The laser beam power is *1 mW* and all the light returning to the lens is focussed onto a detector. How many *photons/second* strike the detector, assuming the laser beam is striking a white perfectly diffuse reflective bar at normal incidence ?

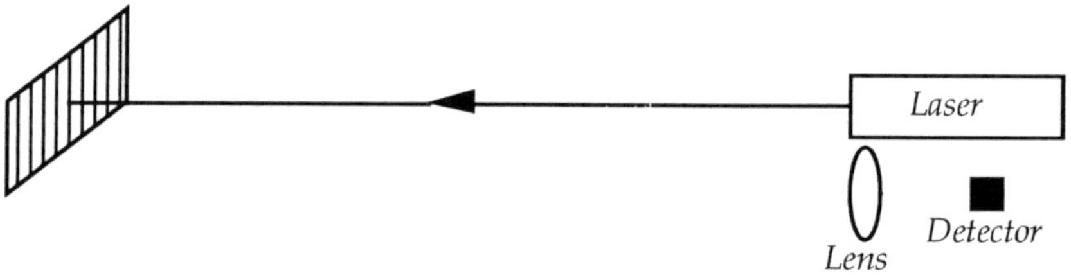

Chapter 2

Interaction of Radiation with Matter: Absorption, Emission, and Lasers

In the preceding chapter we treated quantitatively the two major sources of incoherent radiation, thermal emitters and diffuse scatterers. Our treatment was very general, invoking only thermodynamic and statistical laws, and describing matter in terms of its absorptivity or reflectivity. We now treat in much more detail the interaction of radiation with matter, in particular, the absorption or emission of a *photon*, which we again specify as an increment of energy, $h\nu$, extracted or added to an electromagnetic field of frequency, ν. Einstein, in a classic (but not classical) paper, *(Einstein, 1917)*, placed an assembly of particles in equilibrium with the Planck radiation field and deduced a fundamental relationship for the *induced* and *spontaneous* transitions between the discrete energy levels of the particles. His treatment proved that radiation could *induce* or *stimulate* a *downward* transition, that is, one that resulted in the *emission* of a photon of energy. From this theory, we derive the requirements for net *stimulated emission* or *amplification* in a medium. With an appropriate regeneration or feedback mechanism, such as a Fabry-Perot resonator, we may then obtain *laser* action, or the production of a single-frequency, single-plane-wave electromagnetic field. We treat the Fabry-Perot and discuss the overall laser family after a discussion of Fermi-Dirac or electron statistics, the latter topic being essential for our treatment of semiconductor lasers in Chapter 3.

2.1 The Einstein *A* and *B* Coefficients and Stimulated Emission

We consider the interaction of blackbody radiation with a simple assemblage of particles each of which has two possible energy levels, U_1 and U_2. These particles might be, for example, ammonia molecules, NH_3, which were used in the first demonstration of stimulated emission, the *maser*, a *m*icrowave device. As shown in Figure 2.1, the particles

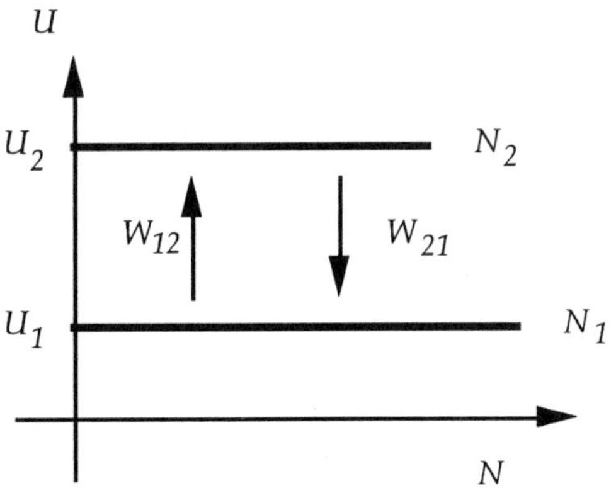

Figure 2.1 Transition rates and occupancy for states in equilibrium with blackbody field.

are distributed such that N_2 are in the upper state and N_1 in the lower state. The particles are contained in a blackbody chamber at temperature T, and the occupancy probability is given by Eq. (1.1), the Boltzmann relation. Thus the ratio N_2/N_1 becomes $e^{-h\nu/kT}$, since energy absorbing or emitting transitions between the two levels occur with a change in electromagnetic field energy of $(U_2 - U_1) = h\nu$. Defining the transition *probabilities per unit time* as W_{12} and W_{21}, we may write that $N_1W_{12} = N_2W_{21}$ since the total rates must be equal in thermal equilibrium.

Now Einstein hypothesized that there were both *spontaneous* and *induced* or *stimulated* transitions. The *spontaneous* transitions occurred from the upper to the lower state with a probability per unit time of A, characteristic of the particle. In contrast the *induced* transition probability was assumed to be proportional to the electromagnetic spectral energy density, u_ν, of Eq. (1.12), and *transitions were induced in both directions*. It was this latter conjecture that was most surprising since absorption had always been treated as the single process of excitation from the lower to the higher energy state. As we will see, it is essential to include the induced *downward* transitions to satisfy the rate equations, which we write as

$$N_1W_{12} = N_1B_{12}u_\nu = N_2W_{21} = N_2(A + B_{21}u_\nu) \tag{2.1}$$

with B_{12} and B_{21} the proportionality constants for the upward and downward induced or stimulated transitions. Manipulation of the second and fourth terms of Eq. (2.1) yields

$$u_\nu = \frac{A}{\frac{N_1}{N_2}B_{12} - B_{21}} = \frac{A}{B_{12}e^{h\nu/kT} - B_{21}} \tag{2.2}$$

But we know from Eq. (1.12) that

$$u_\nu = \frac{8\pi h\nu^3}{c^3(e^{h\nu/kT} - 1)}$$

and, therefore, to satisfy the relationship of Eq. (2.1) for all frequencies and all temperatures, $B_{12} = B_{21} = B$, and $A/B = 8\pi h\nu^3/c^3$. Since we have established that the upward and downward induced rates are equal , we shall use the single constant B, which can be written

$$B = \frac{Ac^3}{8\pi h\nu^3} = \frac{c^3}{8\pi h\nu^3 t_s} = \frac{\lambda^3}{8\pi h t_s} \tag{2.3}$$

where t_s is defined as the *spontaneous emission time* and is equal to $1/A$. The actual values of the A and B coefficients are determined by the specific system considered. Obviously the *larger* the thermal equilibrium absorption, as determined by the B coefficient, the *smaller* the radiative or spontaneous emission time. Also, the higher the optical frequency, the shorter is t_s for the same value of B or absorption coefficient.

2.2 Absorption or Amplification of Optical Waves

Since we are interested in the absorption or emission of a single-frequency or *monochromatic* plane wave , we now propagate such a wave as shown in Figure 2.2. We assume that the intensity is small enough so that it does not disturb the state populations and that the frequency of the wave is ν, which is at or near the "resonant" or characteristic fre-

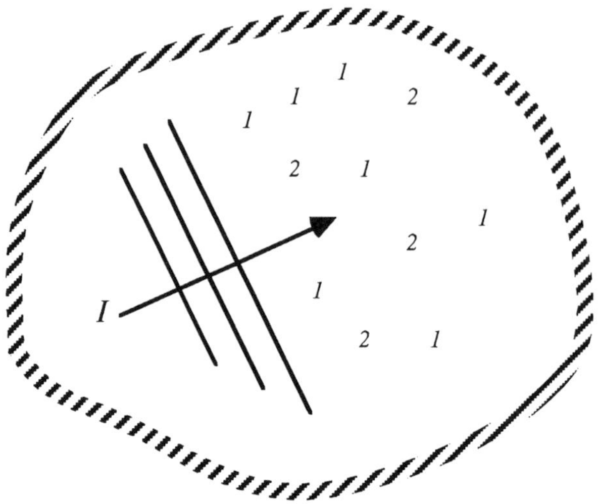

Figure 2.2 Blackbody chamber with added plane wave of intensity *I*. The small numerals represent particles in the upper, *2*, or lower, *1*, state.

quency of the transition, which we now call ν_0. Now as the wave propagates it will induce upward and downward transitions and a concomitant decrease or increase in photons. If we imagine a thin slab of area A and thickness dz the incident power, $P = IA$, will increase by $d(IA) = AdI$, as shown in the figure. Then $dP = AdI = dU/dt = V(du/dt) = Adz(du/dt)$. since the increase in power is equal to the added energy U per unit time in the small volume.

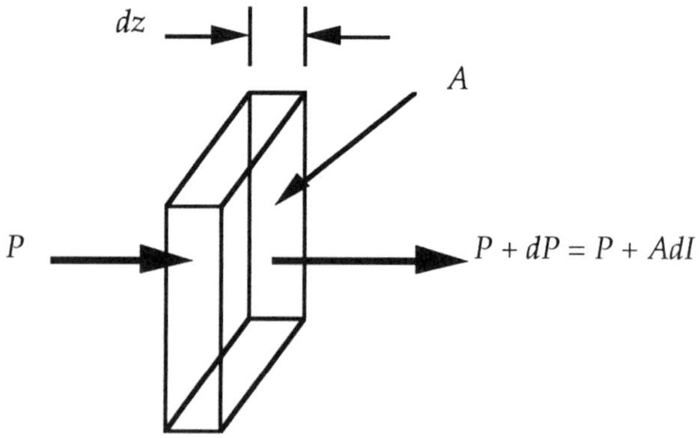

Therefore,

$$\frac{dI}{dz}=\frac{du}{dt}=\frac{1}{V}\frac{dU}{dt}=\frac{1}{V}\left[(N_2-N_1)Bu_\nu\right]\cdot h\nu=h\nu B(n_2-n_1)u_\nu \tag{2.4}$$

where n_1 and n_2 are the number of particles per unit volume. The bracketed quantity in the fourth term is the rate of increase in photons in the volume. But this equation has the interesting problem that the quantity u_ν, the spectral energy density, is actually infinite since we are talking about a *monochromatic* or single-frequency wave. We resolve this dilemma by realizing that energy transitions are allowed over a small range of frequencies about the center or resonant frequency ν_o. Thus the effective spectral energy density u_ν may then be written as $u_\nu = ug(\nu) = Ig(\nu)/c$. The quantity, $g(\nu)$, is known as the lineshape function and is sketched in Figure 2.3. It has the dimensions of inverse frequency and the area under the curve is equal to unity, thus $\int g(\nu)d\nu = 1$. In our original derivation of the *A* and *B* coefficients we used a continuum of frequencies for the electromagnetic energy density with a single-frequency particle transition, while here we have taken into account the finite linewidth of the transition and used a single-frequency wave. The lineshape function describes the finite spread in the emitted spectrum caused by the finite spontaneous emission time and also gives the proportionate response of the induced process as a function of the incident

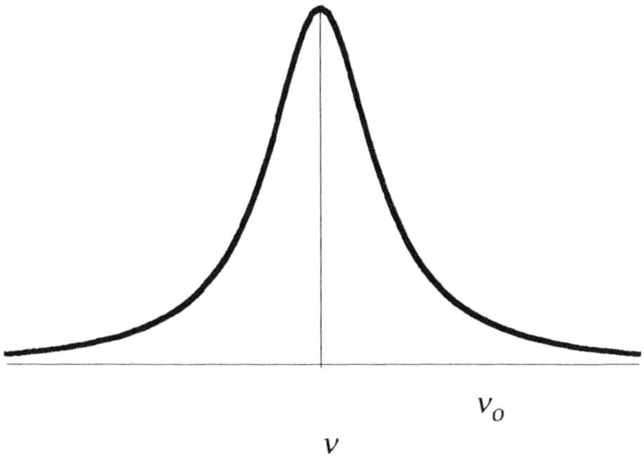

Figure 2.3 The lineshape function $g(\nu)$.

frequency. Equation (2.4) thus becomes finally

$$\frac{dI}{dz} = \frac{I}{c} h\nu \, Bg(\nu)(n_2 - n_1) \tag{2.5}$$

The lineshape function shown in Figure 2.3 is called *Lorentzian* and is identical to the power response curve of a *single-pole* electrical network with a decay time equal to the spontaneous emission time. This same lineshape applies to a system where collisions of the active particle determine the radiative lifetime. In a gaseous system, the line is said to be *pressure broadened*. Another common lineshape is the Gaussian, which is associated with the velocity distribution of the particles and the resultant *Doppler* shift of the resonant frequencies. We speak of such a system as *velocity broadened*. In all cases, the linewidth $\Delta\nu$ is given by $1/g(\nu_o)$. The broadening mechanisms are discussed in *(Yariv, 1991, Ch. 5)*.

Equation (2.5) makes an inherent assumption that we have not justified. Namely, we have assumed that each "photon" or increment of energy added to the wave produces an electric field of the same frequency, phase, and direction as the incident wave. This is rather surprising but can be explained rigorously by a quantum-mechanical treatment. Alternatively, we can consider the absorption process depicted in Figure 2.2. Here the wave induces upward transitions or absorption in lower state species *1*, while inducing emission of extra energy from species 2. In our example, we know that there is a net absorption since there are more *1's* than *2's*. But if the *2's* emitted at different frequency, phase, or direction , we would observe scatter or frequency shifts in the resultant transmitted radiation. This is not observed in an absorbing medium and thus the coherent replication and amplification of the wave for n_2 greater than n_1 is consistent with our simple argument.

Returning to Eq. (2.4), we now calculate the gain (or absorption) coefficient by writing

$$\frac{dI}{dz} = \gamma I \qquad \Rightarrow \qquad I = I(z=0)e^{\gamma z}$$

and γ is given by

$$\gamma = (n_2 - n_1)B\frac{h\nu}{c}g(\nu) = (n_2 - n_1)\frac{c^2 g(\nu)}{8\pi\nu^2 t_s} = (n_2 - n_1)\frac{\lambda^2 g(\nu)}{8\pi t_s} \tag{2.6}$$

using Eq. (2.3) Thus under thermal equilibrium conditions, where n_2 is always less than n_1, γ is always negative, the wave is attenuated and we obtain absorption. If, however, we can somehow "invert" the population, then we can obtain gain or amplification. This process is known as *laser* action from the acronym, *l*ight *a*mplification by *s*timulated *e*mission of *r*adiation. We will discuss lasers in general shortly and then treat the semiconductor laser in detail.

2.3 Electron or Fermi-Dirac Statistics

To understand the semiconductor laser we shall need another new form of thermal statistics, those of electrons, known as Fermi-Dirac statistics. We introduce them here to stress the difference between the simple independent particles we have discussed and the more complicated state occupancy rules for electrons. Rather than allowing any number of independent particles to exist in various energy states U_n, we find that in a solid, there is a finite number of allowed energy states per unit volume per unit energy. In addition each state has an occupation probability equal to one or less. No more than one electron of each *spin* *(±1/2)* is allowed because of the *exclusion* principle. The details of these and the Boltzmann statistics are discussed in *(Reif, 1965)* The probability of occupation of an allowed electron state at energy U is given by

$$p(U) = \frac{1}{e^{(U-U_F)/kT} + 1} \Rightarrow Ae^{-U/kT} \text{ for } \quad U >> U_F; p << 1 \tag{2.7}$$

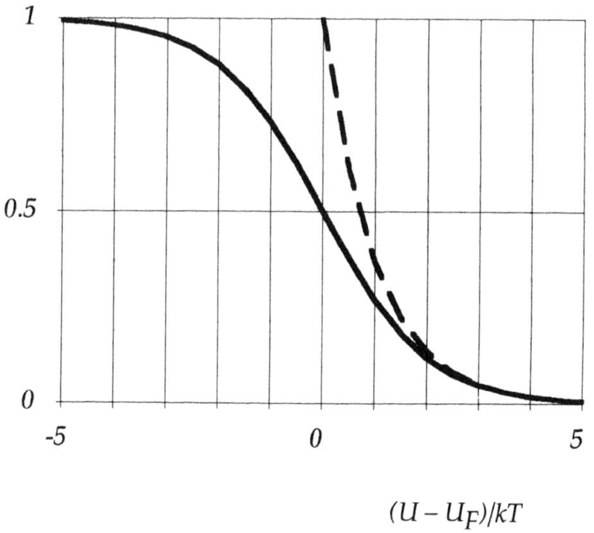

Figure 2.4 Fermi-Dirac (solid) and Boltzmann (dashed) occupancy probabilities.

as shown in Figure 2.4 along with the Boltzmann probability. The quantity, U_F, is called the Fermi energy and is determined by the total number of electrons in the system and the availability of electron states. The Fermi energy is that value of energy which allows all of the available electrons to occupy the given state distribution. We discuss these matters in great detail in the next chapter, but in the meantime, we note that the occupation probability goes to unity at low energies, falls to zero for high energy with the same behavior as the Boltzmann law, and equals one-half at $U = U_F$. In the problems, you are asked to perform the same rate calculations as we did with the Boltzmann distribution, to confirm that the A and B coefficient values are still the same. To perform this exercise, we need the same rules as will be used in the derivation of the gain in semiconductor lasers. These rules are that the transition rate between states of limited occupancy are proportional not only to the occupancy of the initial state but to the "emptiness" or availability of the final state. Thus W_{12} and W_{21} include products of the form $p(1-p)$.

2.4 The Fabry-Perot Resonator

Stimulated emission was first observed at microwave frequencies by Gordon, Zeiger, and Townes in *1955 (Gordon et al., 1955)* They used a beam of ammonia (NH_3) molecules which entered a microwave cavity tuned to approximately *20 GHz*. This resonant frequency corresponds to a transition of the molecule from an upper to a lower vibrational state. The higher energy state has an energy that increases with applied electric field, while the lower state energy decreases with field. By sending the beam of particles along the axis of a "sorter", consisting of a quadrupole electric field in which the field increases with radius, the low-energy particles are deflected and only particles in the upper state enter the cavity. As a result, there is an inverted distribution, the initial molecules emit spontaneously, the radiated field is amplified and the result is an oscillator of exceptional frequency purity. Later microwave masers (*m*icrowave *a*mplification by *s*timulated *e*mission of *r*adiation) used solids as well as gases and are excellent frequency references as well as low-noise microwave amplifiers. In 1958 Schawlow and Townes published a paper discussing the possibilities of "maser" action at optical and infrared frequencies. *(Schawlow and Townes, 1958)* Although there were possible methods of obtaining population inversion, the short spontaneous emission time, inversely proportional to the cube of the frequency, was one of the difficulties to be surmounted. In addition, a "cavity" at optical frequency would be miniscule and alternative structures were needed. Schawlow and Townes proposed the use of a Fabry-Perot etalon, a filter consisting of two highly reflecting, partially silvered mirrors, which has narrow transmission peaks at wavelengths where the spacing is an integral number of half-wavelengths. It was with this structure that laser action was first demonstrated in ruby by Maiman *(Maiman, 1960)*and most laser oscillators use the Fabry-Perot structure or a modified form.

Figure 2.5 shows two partially transmitting mirrors with *amplitude* reflectivity, r, and transmissivity, t. The incident complex field E_i and the transmitted field E_t obey the general relation, $E(t) = Re\,[Ee^{-j\omega t}]$. A wave crossing the cavity in either direction experiences a change in amplitude and phase given by $e^{\delta d}$, where $\delta = -jk + \gamma/2 - \alpha/2$. Here k is the wave vector, $2\pi/\lambda$, γ is the power gain coefficient [Eq. (2.5)], and α is the power attenuation coefficient in the laser material, independent of the gain. The factor of one-half in the δ expression is required since we are considering a field amplitude that is proportional to the square root of the power or intensity. Using the multiple reflections as indicated in the figure we may write the transmitted electric fields at the successive outputs 2, 6, 10, *etc.* as

$$\begin{aligned} E_2 &= t_1 t_2 e^{\delta d} E_i \\ E_6 &= t_1 e^{\delta d} r_2 e^{\delta d} r_1 e^{\delta d} t_2 \\ E_{10} &= t_1 e^{\delta d} (r_2 e^{\delta d} r_1 e^{\delta d})^2 t_2 \qquad \textit{etc.} \end{aligned} \tag{2.8}$$

This leads to an infinite series and a final expression given by

$$\begin{aligned} E_t &= t_1 t_2 e^{\delta d} [1 + r_1 r_2 e^{2\delta d} + (r_1 r_2 e^{2\delta d})^2 + \dots \\ &= \frac{t_1 t_2 e^{\delta d}}{1 - r_1 r_2 e^{2\delta d}} E_i \end{aligned} \tag{2.9}$$

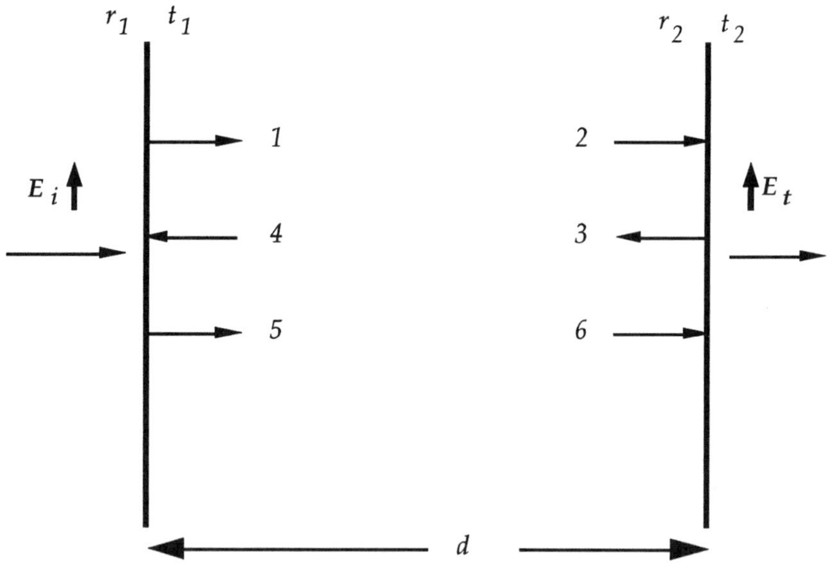

Figure 2.5 Fabry-Perot schematic. Numerals indicate successive passages of a multiply reflected wave.

If we wish to observe oscillation the net transmission must become infinite or the denominator of Eq. (2.8) must become zero. We thus obtain

$$
\begin{aligned}
r_1 r_2 e^{2\delta d} &= 1 \\
e^{2jkd} &= +1 \qquad \therefore kd = n\pi \\
r_1 r_2 e^{\gamma-\alpha} &= +1 \qquad \therefore \gamma = \alpha - \frac{ln(r_1 r_2)}{d}
\end{aligned}
\tag{2.10}
$$

with n a positive integer. Taking into account the expression for k, we obtain the two key equations governing the Fabry-Perot resonator. The first yields the resonant frequencies and the consequent mode spacings; the second establishes the threshold gain, γ_t, required for laser oscillation. Defining the *power* reflectance as $R = r^2$, we obtain finally

$$
\begin{aligned}
&\nu_n = nc/2d \qquad \Delta\nu = c/2d \\
&\gamma_t = \alpha - \frac{ln(R_1 R_2)}{2d}
\end{aligned}
\tag{2.11}
$$

We will use all of these relations in our treatment of semiconductor lasers in the next chapter.

2.5 Inversion Techniques

Three techniques are commonly utilized to obtain an inverted energy-level population in a laser medium. These are optical "pumping," electric discharge energy transfer, and semiconductor electron injection. We leave the last for the next chapter, but here we briefly describe the first two techniques.

Optical pumping, the inversion technique used in the original ruby laser, is an extension of the "three-level" maser scheme proposed by Bloembergen *(Bloembergen, 1956)* for microwave solid-state masers. In Figure 2.6 we show a system of three energy levels, whose occupation, as shown by the length of the bars, obeys the Boltzmann distribution. We now apply a "pump" or intense optical field at the frequency corresponding to transitions between the first and third energy levels. If the energy density is sufficient the level populations will equalize , that is,

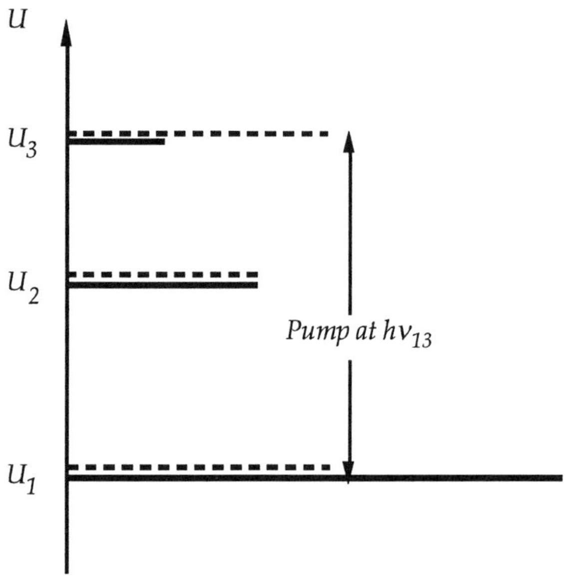

Figure 2.6 Energy level diagram showing optical pumping. The dashed lines showing the final distributions have been displaced for clarity.

$N_1 = N_3$, and in this case N_3 will become greater than N_2, and the *2-3* system will be inverted. The equalization of the *1-3* system results from the pump-induced transitions completely overcoming the spontaneous and thermal radiation-induced transitions. This picture is of course vastly oversimplified, since there are, first, competing nonradiative transitions between levels and, second, the pump source may have a broad spectrum that induces transitions among other level spacings. In most systems, in fact, only the lowest or ground state is occupied at room temperature and the inversion of an upper pair of states is critically dependent on the energies of the states, the pump spectrum, and the relaxation or transition times between states. Albeit, laser action was first demonstrated by Maiman, using a photographic flashlamp and a cylindrical rod of ruby with partially reflecting silvered coatings on the parallel endfaces *(Maiman, 1960)*.

Example: Assume the energy levels in Figure 2.5 are equally spaced, $U_2 - U_1 = U_3 - U_2 = kT$, and the measured *absorption* coefficient for the 2-3 transition is $1\ cm^{-1}$. The system is now "pumped" such that $N_1 = N_3$ and N_2 remains fixed. We now wish to find the *gain* coefficient for the 2-3 transition, assuming the *B* coefficient for the 1-2 transition (at the same frequency) is negligible.

The gain is proportional to (N_3-N_2). The initial value of $(N_3 - N_2)$ is given by $N_2(e^{-1} - 1) = -0.63N_2$. The final value of N_3 is one-half the sum of the initial values of N_1 and N_3 which is $0.5N_2(e^1 + e^{-1})$. Thus the final value of $(N_3 - N_2)$ becomes $N_2[0.5(e^1 + e^{-1}) - 1] = 0.54N_2$. Finally, the *gain* becomes $-1\ cm^{-1}$ times the ratio of the final to the initial value of $(N_3 - N_2) = 0.54/(-0.63) = -0.86$, and is therefore $+0.86\ cm^{-1}$.

The second common inversion technique involves the preferential transfer of energy between and among electrons, atoms, and molecules in an electrical discharge in a gas. The most common example of this type is the helium-neon or He-Ne laser. Here, electron-excited helium atoms transfer energy to an upper state of neon, which ends up with a population much larger than several lower intermediate states. The result is laser action at wavelengths of 3.39 and $1.15\ \mu m$ and $633\ nm$ or $6328\ A$. Another well-developed gas laser is the carbon dioxide system operating at wavelengths near $10\ \mu m$ and yielding continuous powers as high as *50 kW!* The resonator mirror spacing in these devices is usually many centimeters and alignment and diffraction losses can become serious. As a result, most if not all gas lasers use concave reflectors in the so-called confocal or modified confocal configuration. This technique confines the mode laterally as well as longitudinally and helps reduce diffraction losses as well as assure single-frequency-mode operation. Yariv gives a detailed discussion of these resonators as well as an in-depth analysis of the various laser systems *(Yariv, 1991)*.

Problems

2.1 In Section 2.1 we found the ratio of the Einstein *A* and *B* coefficients by placing a Boltzmann distributed two-level system in equilibrium with the blackbody radiation field. Electrons in a semiconductor do not obey Boltzmann statistics because of the exclusion principle, that is, the maximum occupancy of a state is unity. The correct distribution, called Fermi-Dirac, states that the probability of finding an electron in a state of energy, *U*, is given by

$$p(U) = \frac{1}{e^{(U-U_F)/kT} + 1}$$

where U_F is called the Fermi energy and is adjusted to fill all the states with the total number of available electrons. Now repeat our calculation including the fact that the transition rate is proportional to the occupancy p of the initial state times the "emptiness" $(1 - p)$, of the final state.

2.2 A Fabry-Perot resonator is constructed from GaAs which has an index of refraction of $n = 3.5$. The surface is uncoated and thus has a *power* reflectance of $[(n-1)/(n+1)]^2$ and the length of the cavity is 0.5 *mm*.

(a) At a *vacuum* wavelength of *9000 A (900 nm)*, find the mode spacing both in *angstrom* units and in *GHz*. Note that the velocity of light in the medium is c/n.

(b) Assuming no internal losses in the GaAs ($\alpha = 0$), find the threshold gain coefficient γ_t, in cm^{-1}.

2.3 A molecular gas laser operating in the far-infrared region, $\lambda > 10$ μm, has an energy-level system as shown in the sketch, with the occupancy initially at thermal equilibrium at $T = 300$ K. An intense optical "pump" is applied at the frequency of the U_1 to U_3 transition, equalizing N_1 and N_3. Assuming that $(U_2 - U_1)$ may be chosen by picking the right kind of molecule, what is the *shortest* wavelength at which gain may be observed on the *2* to *1* transition ? Assume $(U_3 - U_1) >> kT$.

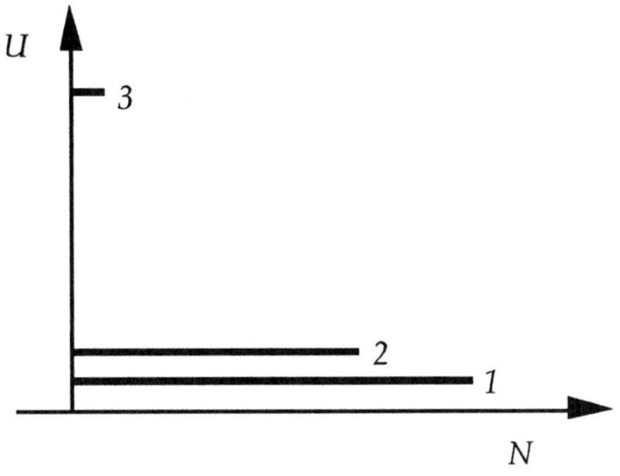

Chapter 3
The Semiconductor Laser

We have now developed a plausible and quantitative description of optical amplification and oscillation and very briefly reviewed inversion techniques and typical lasers. In this chapter we consider in great detail the semiconductor laser. We choose this particular type because of its broad applications, from fiber optic communication to compact disk player readout. We also find our consideration of semiconductor *p-n* junctions useful in a treatment of junction photodiodes in Chapter 5.

Unlike most of the optically pumped laser systems that utilize transitions between discrete energy levels, the semiconductor laser depends on transitions of electrons between a continuum of allowed energies in the *conduction* band and in the *valence* band. In semiconductor device language an electron transition from the higher energy of the conduction band to the lower energy of the valence band is called *electron-hole* recombination. Here we deviate from this simple nomenclature and speak of occupied or unoccupied *electron* states in the valence band. Thus a *hole* in our parlance is an empty or unoccupied electron state in the valence band. We first derive the density of states *versus* energy for a free electron gas, modify the results for application to a semiconductor, and then apply Fermi-Dirac statistics to obtain the thermally controlled electron density *versus* energy. Applying the gain coefficient results from Chapter 2 leads us to an expression for the gain *versus* photon energy as a function of the *injected* electron density. We then discuss typical semiconductor laser structures and calculate the *threshold current* required for laser action and the resultant spectral distribution of the oscillating modes in the Fabry-Perot laser cavity. We close with a calculation of the optical wave behavior from a plane, monochromatic, or *coherent* source, thus obtaining the radiation pattern of the laser output beam.

3.1 Electron State Density *versus* Energy

To understand electron state distributions in a semiconductor, we first treat the simple problem of electrons in a box, sometimes called the *free electron approximation.* The approximation is that the electrons do not interact and that we can neglect questions of charge neutrality in the medium. To calculate the allowed electron states, we utilize a simple relationship from quantum mechanics, that the "wavelength" of the electron is given by $\lambda = h/p_e$ where h is Planck's constant and p_e is the momentum of the electron, given classically by $p_e = mv$. We then say that within a box, such as that shown in Figure 1.2, there are two allowed electron states (one for each of the two allowed electron spins) for each *mode* of the chamber where the modes are now standing waves of the quantum-mechanical electron wave function, given by

$$\Psi = \Psi_o \sin(k_x x)\sin(k_y y)\sin(k_z z)\sin 2\pi \nu t \tag{3.1}$$

where $|\Psi|^2$ gives the probability of finding an electron at the location (x,y,z).

These wave functions are identical to Eq. 1.9 for electromagnetic waves, except that $\nu = U/h$, where U is the electron energy, and $k = 2\pi/\lambda = 2\pi p_e/h$. We may thus count the number of modes or available states in k-space just as we did in Chapter 1, and from the relationship between energy and momentum, determine the distribution of states *versus* energy. Again returning to Chapter 1, we know from Eq. (1.11) that the number of allowed modes, N, is given by

$$dN = \frac{V}{\pi^3}\frac{4\pi k^2 dk}{8} = \frac{V}{2\pi^2}k^2 dk \tag{3.2}$$

Since electrons have positive or negative spin there are two electrons allowed per mode and we can write the final state density in states per unit volume per unit, k, as

$$\rho_k = 2\left(\frac{1}{V}\right)\frac{dN}{dk} = \frac{k^2}{\pi^2}$$

But we want the state density in the energy coordinate, and using the following expressions

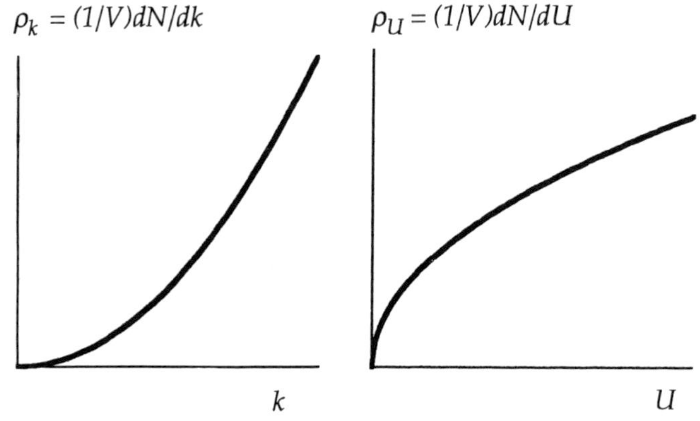

Figure 3.1 Density of states in k-space and U-space.

$$U = p_e^2/2m = h^2k^2/8\pi^2 m \quad and \quad dU = h^2 k dk/4\pi^2 m,$$

we obtain

$$\rho_U dU = \rho_k dk = \frac{k^2}{\pi^2}\cdot\frac{4\pi^2 m dU}{h^2 k} = \frac{1}{2\pi^2}(8\pi^2 m/h^2)^{3/2}U^{1/2}dU \tag{3.3}$$

The typical behavior of the density functions is shown in Figure 3.1. In actual crystals, the density of states behaves somewhat differently, since the electron wave interacts strongly with the lattice structure. For example, if the lattice planes are spaced at a separation of a in the x-*direction*, the periodic potential presented by the lattice of positive ion cores interacts with the electron wave to produce an energy variation with k as shown in Figure 3.2 for the x-component. The simple parabolic behavior of the energy changes drastically in the vicinity of $k = m\pi/a = 2\pi/\lambda$ corresponding to $m\lambda/2 = a$. At this wavelength, the lattice plane spacing produces a resonance similar to that of the Fabry-Perot resonator discussed in Chapter 1. The behavior is the same as the *Bragg* reflection of X-rays in crystals. For such a periodic potential the x-dependent wave function Ψ of Eq. (3.1) becomes, using complex notation,

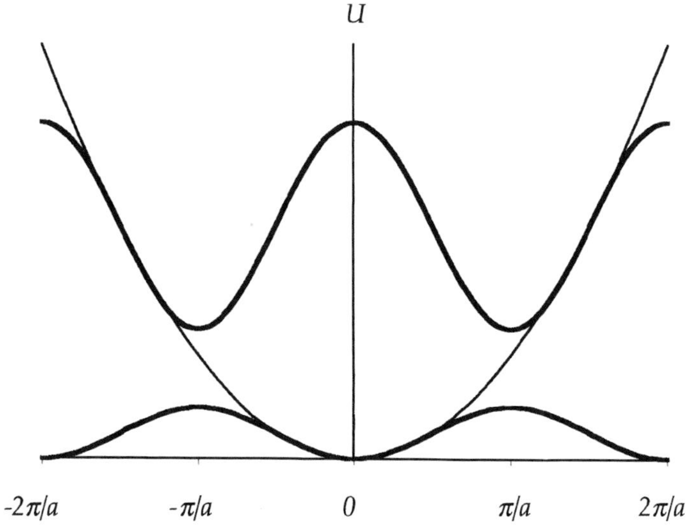

Figure 3.2 Energy U *versus* x-component of wave vector k_x. Light line is empty or uniform potential box. Heavy lines show the effect of periodic potential with lattice plane spacing of a.

$$\underline{\Psi}(x,k_x)=\left[\sum_n \underline{\Psi_{xn}}e^{2jn\pi x/a}\right]e^{-jk_x x} \quad with \quad \Psi(x,t)=Re\left[\underline{\Psi}(x,k_x)e^{j\omega t}\right] \quad (3.4)$$

which is called a *Bloch* wave and here is traveling to the right *(see Callaway, 1991, Ch. 1 or Wang, 1989, Ch. 5)*. The amplitude term in brackets is a Fourier series with fundamental period a. If we now change k_x by $\pm 2\pi m/a$, with m an integer, the wave function remains the same, with a renumbering of the index n in the amplitude series. The energy U is thus periodic in k_x with period $2\pi/a$ as shown in Figure 3.2. Similar relationships hold for the y and z dependence of the wave function. Wave function solutions are beyond the scope of this text, although we can state qualitatively that the amplitude function has a peak value in the plane of the positive ions or where the potential is lowest or most "attractive" to the electrons.

The net result of the modified energy behavior is a density of states as shown in Figure 3.3, where the state density distribution corresponds to the two allowed regions along the dashed line in Figure 3.2. The energies U_C and U_V are, respectively, at the bottom of the conduction

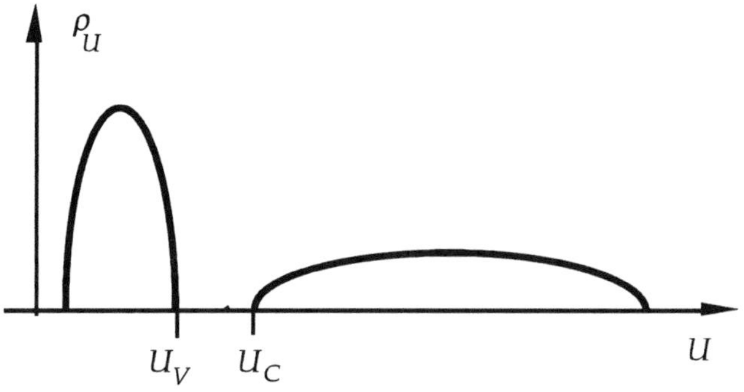

Figure 3.3 State density in a typical semiconductor.

band and the top of the valence band. The variation of $\rho(U)$ with U near the band edges at U_C and U_V is identical to that of Figure 3.1 except for a different curvature and the reverse direction of increasing state density at the edge of the valence band. This results in an "effective" mass in Eq. (3.3) different from that of the free electron. Since the energy of electrons in the valence band *decreases* with increasing momentum or velocity, we sometimes speak of them as having *negative* mass. Consistent with the qualitative behavior in Figure 3.2, the conduction band effective mass m_c is generally smaller than the valence band value m_v. Of course, in device parlance, we treat the absence of an electron as a hole with positive mass.

For an intrinsic semiconductor the total number of available states in the valence band is exactly equal to the number of available electrons. Thus, in the absence of ionized impurities, the band is filled at zero temperature and the material is a perfect insulator. In contrast, a simple metal such as sodium has a completely filled valence band and an extra electron per atom to partially fill the *conduction* band. For the intrinsic semiconductor at finite temperature, the electron distribution is determined by the Fermi-Dirac distribution which we shall rename $f(U)$, to avoid any confusion with p, the hole density, or the density of "empty states" in the valence band. The Fermi-Dirac function plotted in Figure 2.4 may now be multiplied by the density of states function in Figure 3.3, yielding the small density distributions shown at the edges of the energy gap in Figure 3.4. In thermal equilibrium, with the Fermi energy U_F many kT away from the edge of the gap, the occupation probability for

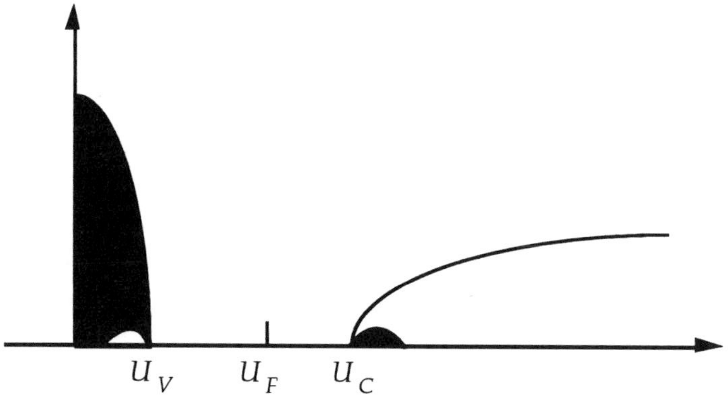

Figure 3.4. Electron density (black) in an intrinsic semiconductor at finite temperature. The hole density is the small white area in the valence band. The Fermi energy U_F is slightly above the center of the energy gap to compensate for the lower state density in the conduction band.

electrons in the conductance band becomes $e^{-(U - U_F)/kT}$, while for electron vacancies or holes, $(1 - f)$ becomes

$$(1-f)=1-\frac{1}{e^{(U-U_F)/kT}+1}=\frac{e^{(U-U_F)/kT}}{e^{(U-U_F)/kT}+1} \Rightarrow e^{-(U_F-U)/kT}$$

for $(U_F - U)/kT >> 1$, since the states are on the "tail" of the Fermi distribution. For gallium arsenide, the original semiconductor laser material, the important constants at room temperature are

$U_g = 1.43\ eV$
m_v *(the hole effective mass)* $= 0.48\ m_e$
m_c *(the electron effective mass)* $= 0.067\ m_e$

where m_e, the free electron mass, is $9.1 \times 10^{-31}\ kgm$.

3.2 Gain Coefficient

To obtain optical gain in a semiconductor material we must obtain a high enough density of electrons in the conduction band and holes or "empty states" in the valence band so that there is a population inversion over some range of optical frequencies. In addition, we must have a reasonable value for the B coefficient in Eq. 2.6 to obtain a high value of gain coefficient γ. This means that the spontaneous or radiative transition time t_s should be small. In gallium arsenide, electron-hole recombination is radiative, and the minority carrier lifetime, of the order of a *nanosecond*, is thus the same as the spontaneous emission time. In contrast, the original bipolar transistor materials, germanium and silicon, have recombination lifetimes from *microseconds* to *milliseconds*. This huge difference is due to differences in the band structure between the two types of material as shown in Figure 3.5. Actual semiconductor crystals have much more complicated U *versus* k behavior than the simple diagram of Figure 3.2. In "indirect gap" material such as germanium or silicon, the lowest conduction band is away from $k = 0$, while in gallium arsenide the lowest band is at $k = 0$. It turns out that radiative transitions between bands occur only for $\Delta k = 0$, or a vertical line in Figure 3.5. Since electrons fill the lowest available conduction band minimum, the recombination in germanium and silicon is indirect. In fact, it is nonradiative and proceeds through intermediate states in the energy gap. The selection rule in k may be explained by conservation of momentum. The initial and final momentum values of the electron are given by $hk/2\pi$ which ranges from 0 to $h/(a/2)$ with a, the lattice con-

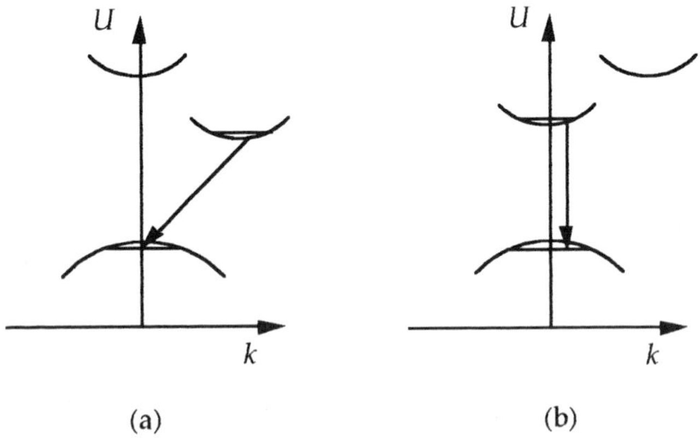

Figure 3.5 Simplified band structure for (a) indirect gap material such as germanium or silicon and (b) direct gap such as gallium arsenide

stant, while the momentum of the emitted photon is h/λ with λ the optical wavelength. Since lattice constants are of the order of *5A* and the optical wavelengths are of the order of *10,000 A*, the allowed change in k is negligible. It is fortunate that the first transistor experiments were performed with germanium whose long lifetime allowed efficient diffusion across the wide base regions of the early structures. It is this same property that makes these materials useless for lasers. Finally, the extra $k \neq 0$ conduction band minimum in gallium arsenide is utilized in *Gunn effect devices* or *TEDs ("transferred electron devices")* as a result of the much lower effective mass of this higher-energy band and the resultant increased mobility *(Wang, 1989, p. 470).*

To this point, we have assumed perfect thermal equilibrium in an intrinsic semiconductor as governed by Fermi-Dirac statistics. It turns out that we can describe a *quasi*-equilibrium in which the conduction band electrons and the valence band holes are separately in equilibrium among themselves. In the presence of excess injected electrons and holes, there will then be a separate Fermi energy for each type of carrier. Thus for a large number of injected carriers, the Fermi energy for electrons moves toward the conduction band, while the hole Fermi energy moves toward the valence band, and the injected electron and hole densities are equal by required charge neutrality. This model's validity is based on the rapid *thermalization* of the carriers due to collisions, since the collision time within the bands, approximately 10^{-13} *sec*, is orders of magnitude smaller than *across-the-gap* or electron-hole interactions. For laser action in a semiconductor, we find it necessary to inject carriers into an active region such that the Fermi energy for electrons, U_{FC} lies above the bottom of the conduction band, U_C, and the Fermi energy for holes, U_{FV}, is below the top of the valence band, U_V. We shall verify this criterion shortly, but first consider the energy diagram of Figure 3.6, where we introduce energies for the initial and final transition states, $U_C(\nu) - U_V(\nu) = h\nu$, and show the Fermi energies for electrons and holes, U_{FC} and U_{FV}. In addition, we show the electron distributions associated with the position of the respective Fermi energies, taken at $T = 0\ K$ for simplicity. The values of the initial and final transition energies depend on the optical frequency and have unique values since the initial and final energy must correspond to the same k vector. Similarly there is a fixed relationship between U_{FC} and U_{FV} since by charge neutrality, there must be equal densities of holes and electrons in

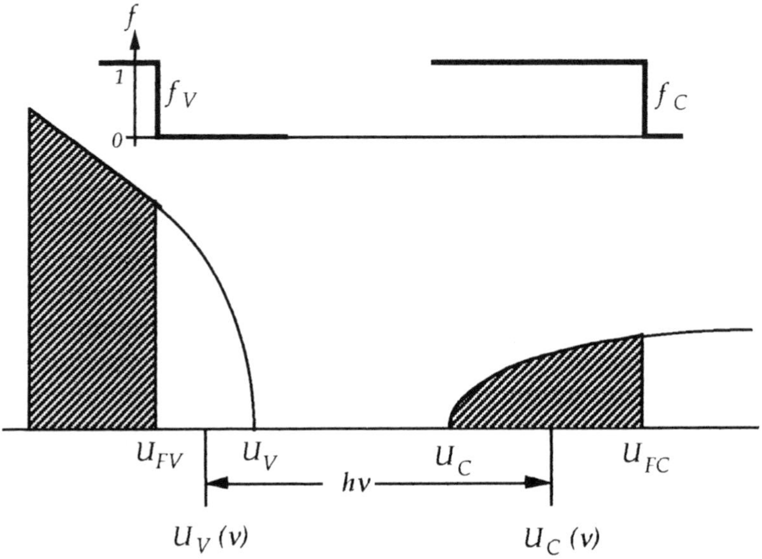

Figure 3.6 Electron and hole densities at $T = 0\ K$ showing position of Fermi energies and the radiative transition, $U_C(\nu) - U_V(\nu) = h\nu$. The Fermi factors, f_V and f_C are sketched at the top.

the interaction region. These densities are determined by integrating the product of the appropriate Fermi function and density of states distribution, and we note that this is a complicated integration for finite temperatures where the Fermi distribution is not a simple step function.

The laser gain coefficient is given in Eq. (2.6) and we now modify this form to calculate the laser gain. This is accomplished by writing the effective value of the upper and lower electron densities, $(n_2 - n_1)$, as

$$\begin{aligned}(n_2 - n_1)_{eff} &= \int \{f_C(\nu)[1 - f_V(\nu)] - f_V(\nu)[1 - f_C]\}\rho_k dk \\ &= \int [f_C(\nu) - f_V(\nu)]\rho_k dk\end{aligned} \tag{3.6}$$

with the Fermi factors f_C and f_V are given by

$$f_C(\nu) = \frac{1}{e^{[U_C(\nu)-U_{FC}]/kT} - 1}; \quad f_V(\nu) = \frac{1}{e^{[U_V(\nu)-U_{FV}]/kT} - 1} \tag{3.7}$$

as shown in Figure 3.6 for $T = 0\ K$. The upper and lower state densities in Eq. (3.6) have each been multiplied by an $f(1 - f)$ term to take into account the "emptiness" of the terminal state, and from Eq. (3.2), the state density ρ_k is the same for each band. The gain from Eq. (2.6) is then

$$\gamma = \int [f_C(\nu) - f_V(\nu)]\rho_k dk \cdot \frac{\lambda^2 g(\nu)}{8\pi t_s} \tag{3.8}$$

But the integral over dk implies a range of k values and thus a range of different frequencies, ν. Actually, the contributions of energy to the single-frequency transition at some ν_o come from a range of k corresponding to the energy spread associated with the linewidth, $\Delta\nu = 1/g(\nu_o)$.

> **Example:** The linewidth $\Delta\nu$ for radiative transitions in a semiconductor is determined by the collision frequency of the electrons which is approximately $10^{13}\ Hz$. The spread in energy of the electrons involved in a single-frequency stimulated transition then becomes $\Delta U = h\Delta\nu = (6.6 \times 10^{-34}) \times 10^{13} \approx 7 \times 10^{-21}\ J$. In terms of electron-volts, this becomes $(7 \times 10^{-21})/(1.6 \times 10^{-19}) \approx 5 \times 10^{-3}\ eV$, an energy spread appreciably smaller than kT and much smaller than the energy range for any significant change in the state density.

Since Δk and $\Delta\nu$ are both small, Eq. (3.8) can be rewritten as

$$\gamma = [f_C(\nu) - f_V(\nu)]\rho_k \Delta k \cdot \frac{\lambda^2 g(\nu)}{8\pi t_s} \tag{3.9}$$

The final form of Eq. (3.9) is obtained by writing all terms as a function of ν and $\Delta\nu$ taking into account the relationships among k, U, and ν, and then using the relation $g(\nu)\Delta\nu = 1$. The result, derived in detail in *(Yariv, 1991, p.559 et seq.)* in slightly different form, becomes

$$\gamma = \frac{\lambda_0{}^2}{2n^2h^2t_s}\left[\frac{2m_cm_v}{m_c+m_v}\right]^{3/2}(h\nu - U_g)^{1/2}[f_c(\nu) - f_v(\nu)] \tag{3.10}$$

where n is the index of refraction of the semiconductor and λ_o is the vacuum wavelength. We note that for positive gain $f_C > f_V$, which from Eq. 3.7 after some manipulation, yields the fundamental laser requirement that $U_{FC} - U_{FV} > h\nu$. This is perfectly consistent with the energy relationships shown in Figure 3.6. As we will see later, this difference in the Fermi levels is equal to the applied forward voltage in the laser diode reduced by any ohmic voltage drop in the bulk semiconductor. Thus the required voltage for laser action is always greater than the photon energy as measured in electron-volts. This is of course a statement of the conservation of energy, since the average electron has to be supplied with an energy greater than that of the emitted photon. Equation (3.10) can be evaluated for gallium arsenide, using the constants of Section 3.1, and $n = 3.5$, $\lambda_o = 0.85\ \mu m$, and $t_s = 3 \times 10^{-9}\ sec$. Here the radiative lifetime, t_s, is the value for a transition to an *empty* valence band state, as deduced from optical absorption data. The wavelength corresponds to a photon energy slightly greater than the energy gap of GaAs, $U_g = 1.43\ eV$. The result is

$$\gamma = 9.7 \times 10^3\left(\frac{h\nu}{q} - \frac{U_g}{q}\right)^{1/2}\left[f_c(\nu) - f_v(\nu)\right]\ cm^{-1} \tag{3.11}$$

which is plotted in Figure 3.7, the heavy curve being for $[f_c - f_v] = 1$. For an intrinsic semiconductor in thermal equilibrium, f_C is effectively zero and f_V is near unity. The result is an absorption coefficient, α, which is the negative of γ and is thus the lowest light curve in Figure (3.7). It is just this absorption curve which gives the experimental data required for the analysis of various semiconductors. Returning to laser action, we must now modify the heavy curve of Figure 3.5 to take into account the f-factors. We know that if the Fermi level is in the conduction band then $(f_C - f_V)$ is a maximum at $h\nu = U_g$ and falls to zero when $h\nu = U_{FC} - U_{FV}$. The net result, shown for two different values of n_{inj}, the electron density in the conduction band, is a range of fre-

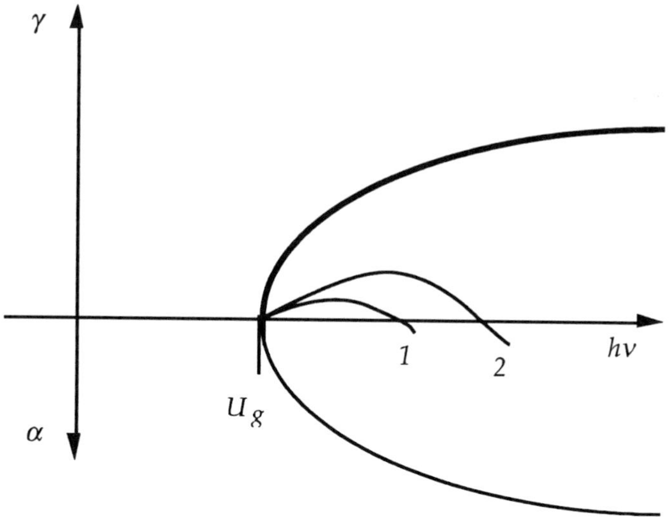

Figure 3.7 Gain as a function of photon energy. The heavy line is for $f_C - f_V = 1$. The lower thin line is for $f_C = 0,\ f_V = 1$ and corresponds to the intrinsic absorption curve. Lines *1* and *2* represent two different levels of injected electron density or values of U_{FC} and the interdependent U_{FV}.

quencies in which the gain is greater than zero. Over a smaller range of frequencies, the gain may be high enough to equal or exceed the threshold gain and laser action may be obtained. We shall consider typical structures and the required threshold gain in the next section. The results of Eq. 3.9 can be combined with the expression

$$n_{inj} = \int_{U_C}^{\infty} f_C(U)\rho(U)dU$$

to obtain the relationship between the maximum gain and a particular electron density. This is a complicated numerical calculation at finite temperature and the results for GaAs are given in *(Yariv, 1991, pp. 563-4)* An empirical expression is

$$\gamma = (1.5 \times 10^{-16})(n_{inj} - n_0) \quad cm^{-1} \qquad n_0 = 1.6x10^{18}\,cm^{-3} \qquad (3.12)$$

where n_o is the electron density at zero gain. Required threshold gains in typical laser structures obtained from Eq. (2.10) are from 20 to 80 cm^{-1} resulting in required electron densities of the order of $10^{18} cm^{-3}$.

3.3 Laser Structures and Their Parameters

Historically, the first semiconductor laser was a *homojunction*, a simple configuration of p- and n-type material as shown in Figure 3.8.

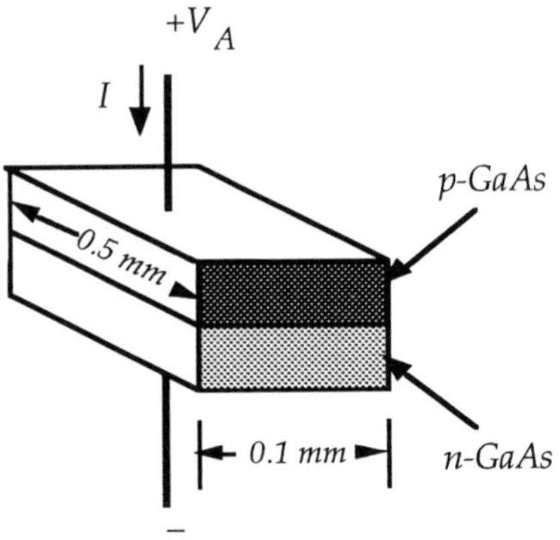

Figure 3.8 Homojunction laser showing typical dimensions.

A Fabry-Perot cavity is formed by the cleaved endfaces, which have a power reflectance of approximately *30%* due to the high index of refraction of the gallium arsenide, $n = 3.5$. Gain occurs in an active region in a thin plane at the junction. Laser oscillation will produce an output beam at the cleaved faces because of the higher overall gain in the long dimension. Figure 3.9 shows the energy band behavior for the diode before and after application of forward bias. The active region at the center of the junction has dimensions controlled by the carrier diffusion length and is typically of the order of $3\ \mu m$.

From Eq. (3.12), the required electron density for threshold or laser oscillation is about $2 \times 10^{18}\ cm^{-3}$, using a typical required threshold gain of approximately $20\ cm^{-1}$ as determined from Eq. (2.11). This leads to a threshold *current* density of

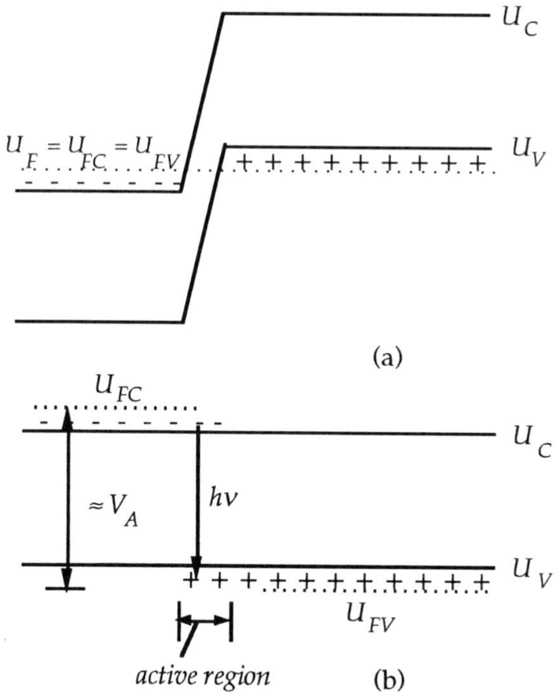

Figure 3.9 The homojunction laser (a) unbiassed and (b) forward-biased.

$$J = \frac{qn_{inj}L}{t_s} = \frac{(1.6 \times 10^{-19})(2 \times 10^{18})(3 \times 10^{-4})}{(3 \times 10^{-9})} = 3.2 \times 10^4\ A/cm^2 \quad (3.1)$$

where L is the diffusion length. This is an extremely high current density and for the structure of Figure 3.8, would result in a threshold current of *16 A*! Consequently, early semiconductor lasers were only operated in the pulsed mode and at liquid nitrogen temperatures *(77 K)*. The low-temperature bath significantly reduces the required electron density and also carries away the excess heat.

The development of the *heterojunction* allowed lasers to be operated continuously at room temperature. The heterojunction consists of dissimilar semiconductor materials with slightly different energy gaps. In a structure such as that shown in Figure 3.10, the carriers are confined to a much smaller region, typically *0.1 μm* in length in contrast to the diffus-

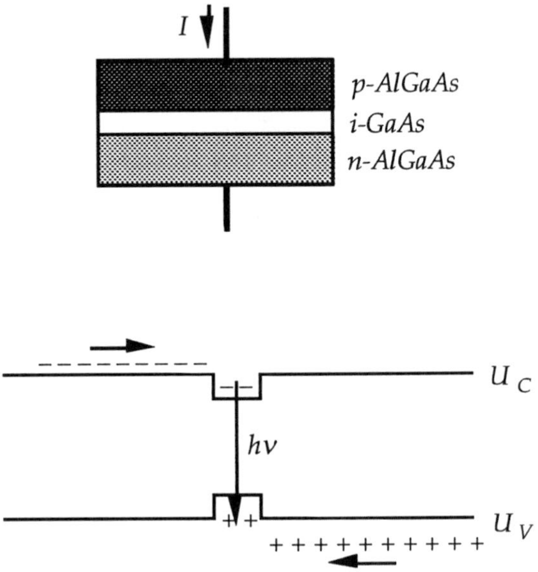

Figure 3.10 Heterojunction composition and energy band structure.

ion-limited length of the homojunction, *3 μm*. Electrons moving to the right and holes moving to the left are trapped or confined to the narrow GaAs region. For a *0.1-μm* active region thickness, the threshold current density becomes $10^3\ A/cm^2$ and the threshold current, *500 mA* This lower current allows operation at room temperature but with an applied voltage of about *1.5 V*, the input power of *0.75 W* is still rather high for continuous operation. This problem can be solved by the use of a "stripe" or "buried" junction configuration as shown in Figure 3.11. The cross-hatched region is the active GaAs while the remainder of the structure is again a ternary such as AlGaAs with a larger gap. In (a), the metallic contact to the *p*-region confines the current to the center of the junction, while in (b) the active area is confined by the insulating material on either side. Using typical dimensions of *3 μm* for the width of the active region, one obtains a threshold current of only *16 ma* because of the reduced area in the direction of the current flow. Since the AlGaAs has a lower index of refraction than GaAs, there is also a confinement of the optical wave to the active region, especially in the buried structure of Figure 3.11(b), where there is a reduced index to the

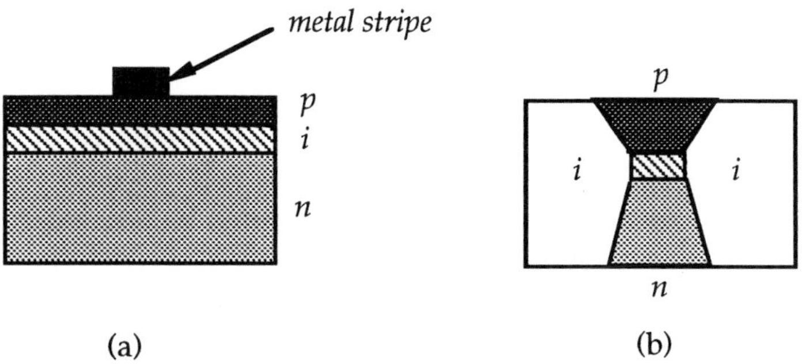

Figure 3.11 (a) "Stripe" geometry and (b) buried heterostructure. The cross-hatched region is the active smaller gap semiconductor.

side as well as above and below. An additional advantage of the heterostructure is the transparency of the material outside the active region; that is, across-the-gap absorption is negligible since the energy-gap of the bounding material is greater than the energy of the laser photons.

Returning to the laser behavior, once the current reaches the threshold value, oscillation commences and the output power increases linearly with input current. Figure 3.12 shows the optical power and electron

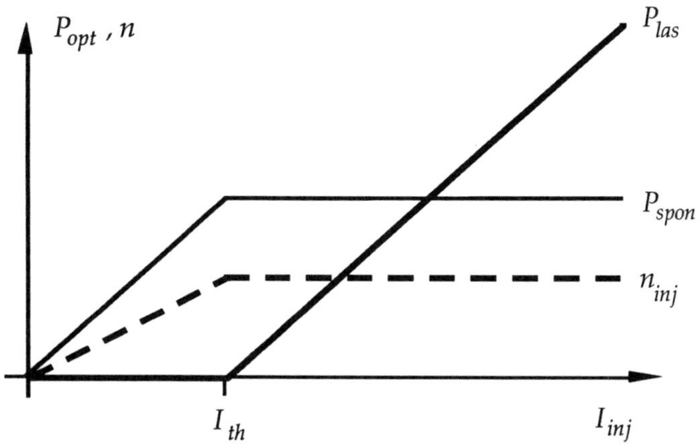

Figure 3.12 Optical output power and injected electron density as a function of injected current.

density behavior below and above threshold. The injected electron (and hole) density and the spontaneous optical power increase linearly with current until the threshold is reached. Then they become constant above threshold as the gain saturates and the recombination time decreases because of stimulated emission. Ideally there would be one photon emitted for each injected electron, however losses due to current leakage, free carrier absorption, and nonradiative electron transitions lead to incremental or "slope" efficiencies of the order of *50%*.

When the laser first starts to oscillate, many of the allowed Fabry-Perot modes are excited and the output beam will contain several different frequencies, as shown in Figure 3.13. If we think of oscillation as the saturated amplification of spontaneous radiation into the many allowed modes, then initially, just above threshold, all modes with net gain greater than zero will produce a small amount of output power approximately proportional to the net gain as shown by the heavy vertical lines in the figure. As the injection current increases, more and more electrons are stimulated to radiate in the strongest mode and eventually one mode becomes dominant and single-frequency operation is obtained. In laser parlance this is described as single-longitudinal-mode operation. In early lasers, or in any structure with a wide active region, such as the homopolar structure of Figure 3.8, there may exist "transverse" modes corresponding to waves moving at a slight deviation from normal at the reflecting endfaces. These modes have frequencies that fall between those of the main longitudinal modes.

Following the original demonstration of laser action in gallium arsen-

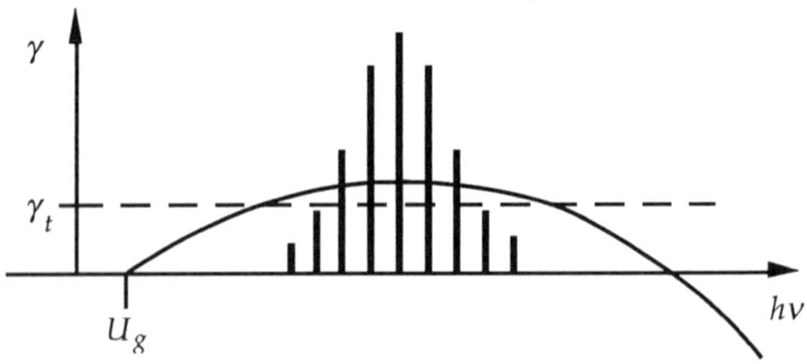

Figure 3.13 Gain curve and threshold gain and resultant mode power distribution.

ide, a large family of lasers has been developed using a variety of semiconductor materials, both in homojunction and heterojunction form. Lasers have been operated at wavelengths as long as *10 μm* using lead tin telluride (PbSnTe) and at the important fiber optic transmission wavelengths, *1.3* and *1.5 μm*, using gallium indium arsenide (GaInAs) and gallium indium arsenide phosphide (GaInAsP). A chart of the available energy gaps in the III-V semiconductor family is given in *(Yariv, 1991, p. 572).*

Finally, as a result of significant advances in fabrication techniques, *vertical cavity* lasers are now realizable as shown schematically in Figure 3.14. The key element in this structure is the multilayer dielectric mirror composed of quarter wave layers of semiconductor with alternating high and low indices of refraction. *(Iga and Koyama, 1993)* In addition, similar fabrication techniques are used to produce *quantum-well* structures in either type of laser. In this case, the active region is thin enough, *50* to *100A* or comparable to the electron wavelength, to produce a marked change in the simple parabolic density of states distribution, and a concomitant additional control over threshold currents and operating wavelengths (*Zory, 1993).*

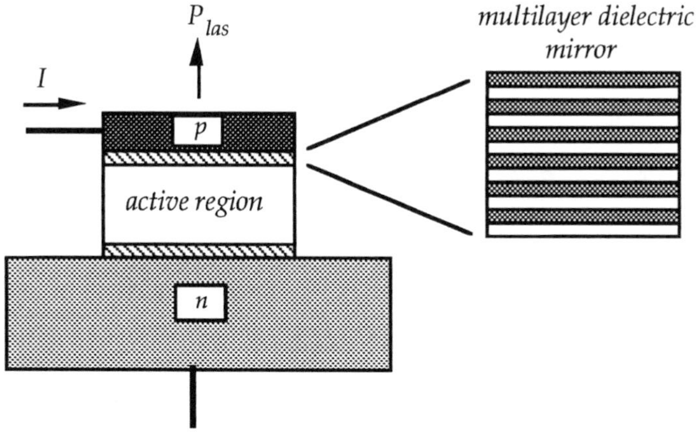

Figure 3.14 Vertical cavity laser showing structure of one of the two multilayer dielectric mirrors.

3.4 Laser Beamwidth and the Fraunhofer Transform

In a laser such as the buried heterostructure type, the optical wave inside the device is guided by the difference in indices of refraction between the active and surrounding regions. Even in the absence of an index step, the high gain tends to confine the wavefront to the vicinity of the active region. Whether "index-guided" or "gain-guided," the beam profile can be reasonably well described by a Gaussian, $\exp(-ax^2 - by^2)$, the coefficients dependent on the height and width of the active region. As the beam emerges from the laser, it propagates through free space, obeying the electromagnetic wave equations, and we derive here what is called the *far-field distribution* produced by such a coherent source.

As shown in Figure 3.15, we ask for the electric field, $\mathbf{E}_F$, at a long distance from a plane containing a *near-field distribution*, $\mathbf{E}_N$. If we define $E(t) = Re[Ee^{j\omega t}]$, then, from radiation theory (*Ramo et al., 1984, p. 611*) we may write the contribution to $\mathbf{E}_F$ due to $\mathbf{E}_N$ as

$$dE_F(x',y') = \frac{j}{\lambda r} E_N(x,y) e^{-jkr} dA \qquad (3.12)$$

where k, the wave vector, equals $2\pi/\lambda$. This expression is called the "paraxial" approximation which assumes that the angle between r and

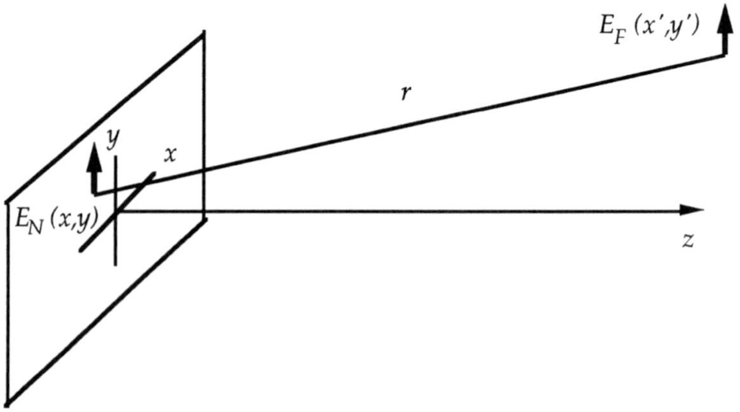

Figure 3.15 Far-field pattern E_F and near-field pattern E_N.

the z-axis is small. For clarity, we first treat only the behavior in the y–z plane as shown in Figure 3.16. We again assume that the distance r is much greater than the extent of the near-field distribution in the y-dimension and that θ_y is small compared to unity. In this case, we can replace r in Eq. (3.12) by the expression $(R - \theta_y y)$, since the distance from the x-y plane to the far-field point decreases by $\theta_y y$ if we move from $y = 0$ to $y = y$. The final value of the far-field then becomes, from Eq. 3.14,

$$dE_F(\theta_x,\theta_y) = \frac{j}{\lambda(R-\theta_x x-\theta_y y)} E_N(x,y)e^{-jk(R-\theta_x x-\theta_y y)}dA$$

Setting $r \approx R$ in the denominator of the coefficient and ignoring the j, which is an optical quarter-wave phase shift, we obtain

$$E_F(k_x,k_y) = \frac{e^{-jkR}}{\lambda R}\iint E_N(x,y)e^{jk_x x+jk_y y}dxdy \tag{3.15}$$

where the angles in the x- and y-directions have been replaced by the x- and y-components of the k-vector, $k_x = k\theta_x$ and $k_y = k\theta_y$. This expression is the Fraunhofer transform and is seen to be a two-dimensional Fourier transform, k replacing ω, and x or y replacing t in the more familiar usage. Thus, for example, the far-field angular pat-

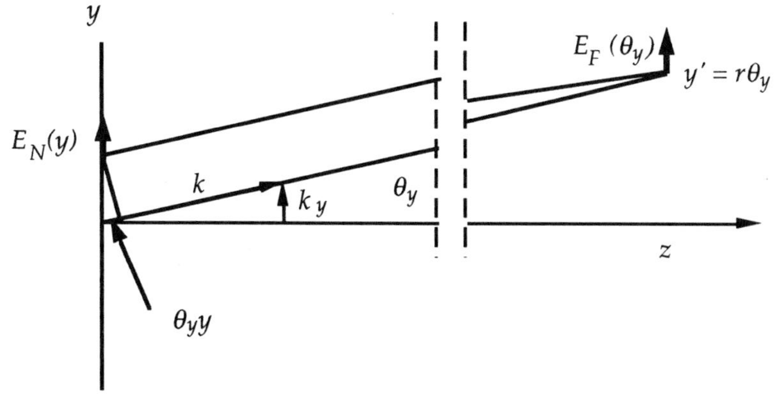

Figure 3.16 Construction for calculating far-field pattern. The two vertical dashed lines represent a large distance between the y and y' coordinate frame.

tern from a slit, which corresponds to a square pulse in the time domain, is a $sin(a\theta)/(a\theta)$, the same as the frequency spectrum of the pulse.

We use two important near-field/far-field relationships in our analyses. The first is the far-field pattern of a uniform circular near-field excitation such as that produced by a plane wave radiated from a circular aperture. The second is the pattern produced by a Gaussian field distribution, which turns out to be also a Gaussian in the far field.

First consider a uniform distribution of constant amplitude E_0, bounded by a circle of radius, a. The transform can be calculated using Eq. (3.15) after conversion to circular coordinates. The result includes a Bessel function of first order and is given by

$$E_F = E_0 \frac{\pi a^2}{\lambda R} e^{-jkR} \left(\frac{2J_1\left[\frac{2\pi a\theta}{\lambda}\right]}{\left[\frac{2\pi a\theta}{\lambda}\right]} \right) \tag{3.16}$$

where the near and far fields are given as scalars since they are each the same component of polarization. This distribution is similar to the $sinx/x$ solution for a one-dimensional square pulse except that the variation is now in the radial direction and the zeros occur at the zeros of the Bessel function $J_1(x)$. The field amplitude is shown in normalized form

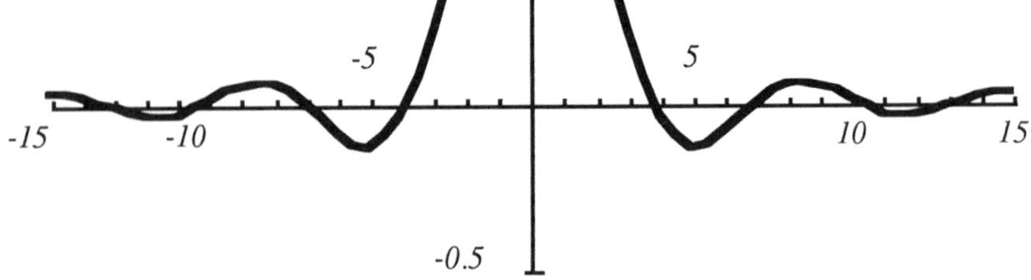

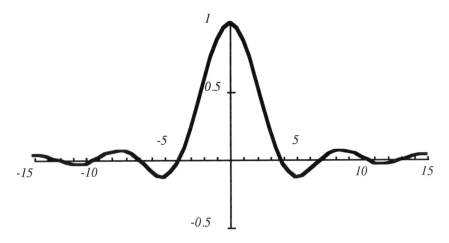

Figure 3.17 The far-field distribution of a circular aperture of radius, a. The normalized amplitude is given as a function of the variable, $2\pi a\theta/\lambda$.

in Figure 3.17, and the intensity, the square of the function, often called an Airy pattern, has a bright central disk surrounded by a series of rings. The pattern is often seen in the telescope image of a point star, since as we shall see, the field pattern at a lens plane acts as the near-field source for a far-field pattern at the focus of the lens. Since the first zero in the function occurs at $2\pi a\theta / \lambda = 3.83$, the angular *radius* of the "Airy disk" becomes $\theta = (3.83/2\pi)(\lambda/a) = 1.22(\lambda/a)$. Another convenient relation is the angular *diameter* of the beam at one-half its maximum intensity or the "half-power beamwidth". This may be found from Figure 3.17 to be very close to the quantity $\lambda/2a$ or λ/d where d is the aperture diameter.

Since most lasers, either with concave mirrors or with dielectric guiding, have a near-field pattern that is close to a Gaussian shape, we now consider the Fraunhofer transform of a field pattern of the form

$$E_N(x,y) = E_0 e^{-ax^2 - by^2} \tag{3.17}$$

which has a far-field pattern, using Eq. (3.13) of

$$E_F(k_x, k_y) = \frac{E_0}{\lambda R}\sqrt{\frac{\pi}{a}}\sqrt{\frac{\pi}{b}}e^{-\frac{k_x^2}{4a}}e^{-\frac{k_y^2}{4b}} \tag{3.18}$$

which is a Gaussian in the angle domain, with $\theta_{x,y} = k_{x,y}/k = \lambda k_{x,y}/2\pi$. We may now write some relationships between the effective near-field area A_{eff} and the effective far-field solid-angle Ω_{eff}. If we define these quantities as follows, we obtain for the Gaussian beam

$$A_{eff} = \frac{\int E_N dA}{E_N(0)} = \frac{\pi}{\sqrt{ab}}; \qquad \Omega_{eff} = \frac{\int E_F d\Omega}{E_F(0)} = \frac{\lambda^2\sqrt{ab}}{\pi} \tag{3.19}$$

and the area-solid-angle product becomes λ^2 and is independent of the size of the original beam. Now in both the near and far fields, contours of constant power are elliptical and we can show that for an ellipse area of $\pi x_o y_o = A_{eff}$, $x_o = 1/\sqrt{a}$ and $y_o = 1/\sqrt{b}$ are the semi-axis distances at which the *power* decreases by e^{-2}. A similar relation holds for the far-field angular pattern and we may write the general relationships

$$\theta_{x0} = \frac{\lambda\sqrt{a}}{\pi} = \frac{\lambda}{\pi x_0}; \quad \theta_{y0} = \frac{\lambda\sqrt{b}}{\pi} = \frac{\lambda}{\pi y_0} \tag{3.20}$$

where the θ's are the e^{-2} *power* locations on the angle semi-axes.

Finally we show that it is not necessary to move a large distance from the source to measure the far-field pattern. Figure 3.18 is a redrawing of Figure 3.16 with the extended region between the dashed lines replaced by a lens. If we think of all field contributions moving at angle θ_y as a rays moving in that direction, we know from simple lens theory that the rays will come to a focus at a position $y' = f\ \theta_y$, as shown in Figure 3.19.

Thus the far-field pattern may be examined by placing the source to the left of the lens and examining the image at the focal point (provided that all rays from the source pass through the lens). For example a Gaussian near-field pattern to the left of the lens with parameters, x_o, y_o, would produce an elliptical intensity pattern on the right focal plane

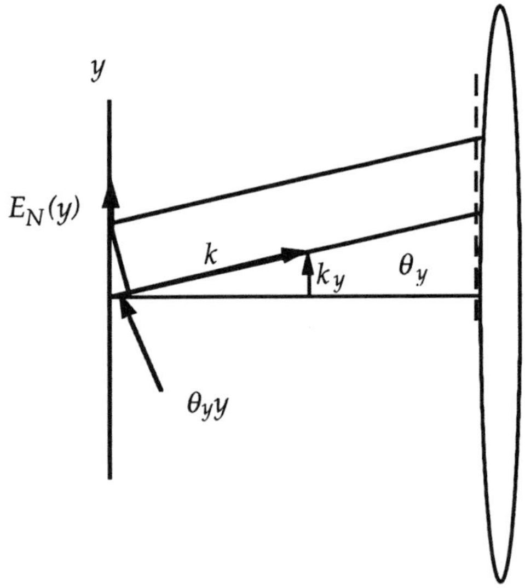

Figure 3.18 The extended region to the right of Figure 3.15 has been replaced with a lens of focal length, *f*.

with parameters, $x',y' = \lambda f/\pi x_0, \lambda f/\pi y_0$, from Eq. (3.20).

In a similar vein, if a uniform plane wave strkes the lens from the left, all the wave power intercepted by the lens will be focused to an "Airy" pattern at the right focal plane, since the "near-field" source is now constant and circular. Since lasers can produce plane or spherical coherent waves, the power may be concentrated in a *diffraction-limited* focal spot, thus making available unprecedented irradiance levels at a surface.

Example: A *300 K* unit emissivity source radiates *460 W/m*^{2} so that a perfect, *f/#* = *0.5*, optical system could produce a focal irradiance of *460 W/m*^{2}. In contrast consider a *10-μm* wavelength *plane* wave of radiance, *460 W/m*^{2}, striking a *1-cm*-diameter lens with a focal length of *5 cm*, or *f/3* = *5*. The power at the focus is then the radiance times the lens area or approximately *40 mW*. The half-power diameter ofthe focal spot is $\lambda f/d = \lambda f/\# = 50\ \mu m$, so the area is approximately $2 \times 10^{-9}\ m^2$. The peak irradiance is then about $(0.04/2 \times 10^{-9}) = 2 \times 10^7\ W/m^2$! This is the same order of magnitude as the radiance of the surface of the sun.

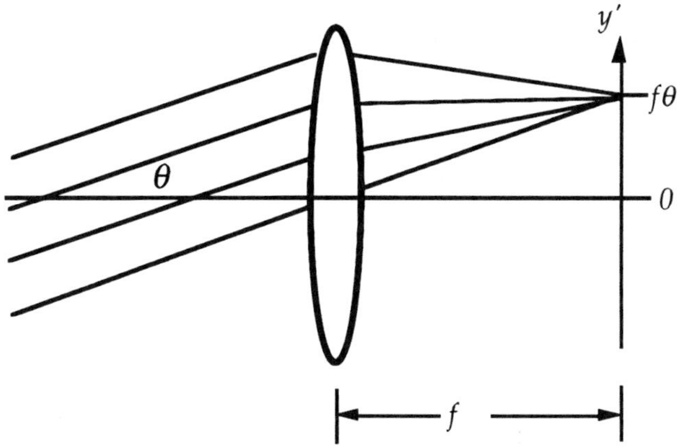

Figure 3.19 A lens converts the far field angle into a position in the image plane

Problems

3.1 A free electron in a solid has an energy of kT at room temperature. Find the wavelength of the quantum mechanical wave associated with this electron. Compare this number with the lattice (atom-to-atom) spacing of a simple cubic crystal with 10^{22} *atoms/cm*^{3}.

3.2 The external quantum efficiency η_{ext} of a semiconductor laser is defined as the fractional number of photons per injected electron. For a laser operating at a vacuum wavelength λ, write an expression for the quantum efficiency in terms of the laser output power P, the wavelength, and the injection current I. Write this expression in numerical form for a wavelength of $1\ \mu m$, with P in *milliwatts* and I in *milliamps*.

3.3 A buried heterostructure laser has an output wave with an effective height of $0.5\ \mu m$ due to the spreading of the wave above and below the active region. The width is $3\ \mu m$ and the laser is emitting $10\ mW$. Find the temperature of a unit emissivity blackbody that could emit the same power per unit area.

3.4 (a) A *1-m*-diameter telescope is used to transmit a CO_2 laser beam to the moon. Calculate the e^{-2} power diameter of the beam spot on the lunar surface, assuming the Gaussian-shaped transmitted beam falls to e^{-2} power at the edge of the telescope. $R = 385{,}000\ km$, $\lambda = 10\ \mu m$.

(b) A semiconductor laser at $\lambda = 0.9\ \mu m$ is focused onto a compact disk recording. The laser beam from the semiconductor overfills the *3-μm*-diameter lens, producing close to a constant intensity over the lens face. If the focal length f *is 1 cm*, find the half-power diameter of the illuminated spot.

Chapter 4
The Ideal Photon Detector and Noise Limitations on Optical Signal Measurement

In the previous chapters we have treated the theory and behavior of the two significant types of optical or infrared source, first, the blackbody or a diffuse reflector, and second the laser. We now consider the optimum means of detection for the radiation from the sources, realizing that our detector, with rare exceptions will be situated at the focus of an optical receiver which in its simplest form is a lens. Historically, the first radiation sensors were thermal detectors, that is, devices whose temperature was a function of the incident optical power, such temperature then being electrically sensed to determine the power level. Actually, Herschel, in an early experiment that established the existence of infrared radiation, used a simple glass thermometer with a blackened reservoir bulb. Current thermal detectors are commonly used where sensitivity and speed of response are not critical and they depend on a temperature-induced voltage (thermocouple), resistance (bolometer), or dielectric polarization (pyroelectric) *(Kingston, 1978, Ch. 7)*. A distinct advantage of the thermal detector is its broad wavelength response. As long as the "blackening" of the element produces appreciable absorption at the wavelength of interest, the detector will respond efficiently. Of course, the detector size must be many wavelengths so that it encompasses the diffraction-limited focal spot of the optical (or infrared) collection system.

In this text, we are interested in the ultimate performance of optical detection systems and will limit our detailed discussions to the "photon" detector, a device that ideally extracts the maximum amount of information available from the incident radiation. Thus far, we have treated in detail both thermal sources as well as coherent or laser sources. Any of these sources produces radiation by either spontaneous or stimulated *downward* energy transitions which release packets of energy, $h\nu$, to the electromagnetic field. We now consider the absorption process and in particular a process where an *upward* transition of an electron absorbs a quantum of energy and we are able to sense the presence of the electron in the higher energy state in an external electrical circuit. If the time-dependent flow of these electrons were an *exact* replica of the time dependence of the incident optical power, we would have a perfect or *noiseless detector.* Unfortunately, the electron production process is not only discrete, one particle at a time, but it is also random, that is, only the *average* rate of electron production reproduces the incident optical power variation and there are extraneous fluctuations which we call noise.

We first treat this phenomenon, known as *shot noise,* for a simple "ideal" detector such as a vacuum photoemitter. If there were some way to measure this shot-noise-limited signal current in a noiseless manner we would have "photon-noise" or signal-noise-limited detection, the noise arising from the discrete and random removal of energy quanta from the radiation field. At this point, however, thermal radiation of the circuit components becomes another source of noise. Specifically, any resistor in the output circuit radiates radio frequency energy just as a blackbody only here the energy is radiated into the *electric circuit.* The *fluctuations* in this radiated energy are called Johnson or thermal noise and we derive the resistor-generated noise current using exactly the same model as that of Chapter 1. We then discuss low-noise amplifiers, such as the *transimpedance* amplifier, which can minimize the effects of electrical circuit noise on the final output signal. Finally we derive expressions for the *signal-to-noise* ratio and *NEP (noise equivalent power)* of detection systems in the signal-noise-, background-noise- and amplifier noise-limited cases.

4.1 The Ideal Photon Detector and Shot Noise

The ideal photon detector can be modeled on the photoelectric effect first analyzed in detail by Einstein. If we shine light on a metal, electrons will be emitted into the surrounding vacuum if the photon energy $h\nu$ is greater than the quantity $q\phi$, where ϕ is called *the work function* and is the added energy in electron-volts needed for an electron of charge q to overcome the potential barrier at the surface. If we now have a positive electrode in the same vacuum envelope, then a current will flow from the illuminated surface, the *photocathode,* to the positive electrode, the *plate.* The process is relatively inefficient because not all the optical energy is absorbed and not all the electrons are moving in the right direction to escape the surface. The important physical property of the process is the probabilistic nature. The induced upward transitions of the electron are determined by the Einstein B coefficient, which we have discussed previously. This rate, however, is not fixed but is a random process which can be described by Poisson statistics. It is this fluctuation in electron emission times which introduces noise into the measurement or signal detection process.

The current induced by incident optical power P can be written

$$\dot{i} = qr = \frac{\eta q P}{h\nu} \tag{4.1}$$

where $r = P/h\nu$ is the average rate of emission of photoelectrons, governed by the quantum efficiency η, which is the fractional number of emitted electrons per incident photon. Although P may be perfectly constant, there is a fluctuation in i because of the random nature of the emission process. If we count the number of photoevents k, during a time τ, when the expectation value or average number $n = r\tau$, Poisson statistics *(see, e.g., Davenport and Root, 1958)* yield

$$p(k,n) = \frac{n^k e^{-n}}{k!}; \qquad \sum_{k=0}^{\infty} p(k,n) = e^{-n} \sum_{k=0}^{\infty} \frac{n^k}{k!} = e^{-n}(e^n) = 1 \tag{4.2}$$

where we have also proven the normalization condition. We use this specific relationship later when we discuss "photon" counting, but for now we note that the mean square fluctuation is $\overline{(k-n)^2} = n$, a familiar relation from statistics.

Example. We will frequently calculate *mean square fluctuations* in the following chapters, so let us verify the preceding value for the Poisson-distributed process. We start with the expression for the mean square fluctuation of a quantity a:

$$\overline{(\Delta a)^2} = \overline{(a-\bar{a})^2} = \overline{a^2 - 2a\bar{a} + (\bar{a})^2} = \overline{a^2} - 2\bar{a}\bar{a} + (\bar{a})^2 = \overline{a^2} - (\bar{a})^2 \tag{4.3}$$

where the overhead bar indicates an average over the distribution and the last term gives us a direct method of calculating the desired result. From Eq. (4.2), the *mean* and *mean square* values of k become

$$n = \bar{k} = \sum_{k=0}^{\infty} kp(k,n) = \sum_{k=0}^{\infty} \frac{kn^k e^{-n}}{k!}$$

$$\overline{k^2} = \sum_{k=0}^{\infty} k^2 p(k,n) = \sum_{k=0}^{\infty} \frac{k^2 n^k e^{-n}}{k!}$$

and we then write

$$\overline{k^2} - n = \sum_{k=0}^{\infty} \frac{(k^2 - k)n^k e^{-n}}{k!} = \sum_{k=0}^{\infty} \frac{k(k-1)n^k e^{-n}}{k(k-1)(k-2)!}$$

$$= \sum_{m=-2}^{\infty} \frac{n^{(m+2)} e^{-n}}{m!} = n^2 \sum_{m=0}^{\infty} \frac{n^m e^{-n}}{m!} = n^2 \quad with \;\; m = (k-2)$$

thus proving that $\overline{(k-n)^2} = \overline{k^2} - n^2 = n$. After the change in index from k to m, the last summation becomes unity from Eq. (4.2). The starting index has been changed from $m = -2$ to $m = 0$, since $(-2)!$ and $(-1)!$ are both infinite resulting in zero for the corresponding terms in the summation.

Since the current during time τ is proportional to k, and the average current is proportional to n, then the *mean-square* fluctuation in the current, which we shall call $\overline{i_n^2} = \overline{(\Delta i)^2}$, is proportional to i. An exact expression for the noise, taking into account the impulse response $i(t)$ produced by both the device and the electric output circuit is called Carson's theorem *(Davenport and Root, 1958)*, which is

$$(\overline{i_n^2})_f = 2r|i(\omega)|^2 \quad with \;\; i(\omega) = \int_{-\infty}^{\infty} i(t) e^{j\omega t} dt \tag{4.4}$$

Here, $(\overline{i_n^2})_f$ is the spectral density or the mean square noise current per unit frequency interval and we are henceforth denoting *electrical* circuit frequencies by f or ω, which has only positive values in our treatment. For frequencies much less than the reciprocal of the pulse width, the magnitude of $i(\omega)$ becomes q, since $i(t)$ becomes a delta function of area q the electron charge. If we define B as the effective power bandwidth of our detection system, then we obtain the widely used *shot noise* expression,

$$\overline{i_n^2} = \int (\overline{i_n^2})_f df = 2rq^2B = 2(\frac{i}{q})q^2B = 2qiB \tag{4.5}$$

An alternative derivation using the expression $\overline{(k-n)^2} = n$ can be found in *(Kingston, 1978)*.

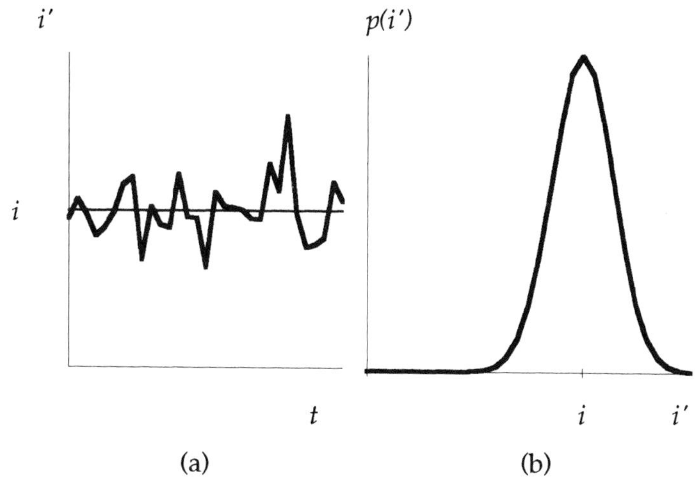

Figure 4.1 (a) Current waveform after addition and low-pass filtering of multiple pulse sequences and (b) current probability distribution.

As a result of this noise, the current output from the detector, instead of being constant for a constant optical power input, fluctuates as shown in Figure 4.1(a). For many individual events per sampling time, or r much greater than B, the distribution of current values about the mean i becomes Gaussian and is given by

$$p(i') = \frac{1}{\sqrt{2\pi \overline{i_n^2}}} e^{-(i'-i)^2/2\overline{i_n^2}} \tag{4.6}$$

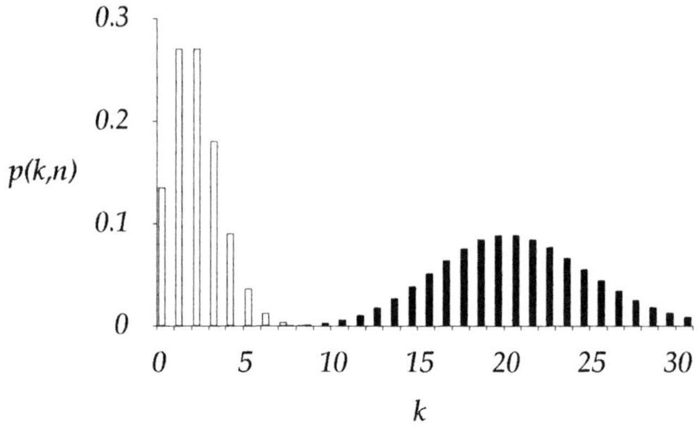

Figure 4.2 Poisson probability distribution. Open bars, $n = 2$; solid bars, $n = 20$.

which is plotted in Figure 4.1(b), with i', the instantaneous value of the current and i, the mean value as determined by the input optical power P.

Example. The Gaussian behavior in the large sample limit is apparent from Figure 4.2 showing the Poisson distribution $p(k,n)$ for $n = 2$ and $n = 20$. For the $n = 20$ case, let us take the measurement time, τ, as $1\ \mu sec$, yielding a current of 20×10^6 *electrons/sec* or 3.2×10^{-12} *A*. A measurement time of $1\ \mu sec$ corresponds to a bandwidth of *500 kHz (from the Nyquist sampling theorem; see also Section 4.5)*, resulting in a mean square noise current of

$$\overline{i_n^2} = 2qiB = 5.12 \times 10^{-25} = q^2 \overline{(k-n)^2} / \tau^2 \quad amp^2$$

from which $\overline{(k-n)^2} = 20$, and the rms fluctuation is $\sqrt{20} = 4.5$ consistent with Figure 4.2.

If the *shot noise* of Eq. (4.5) were the only noise or current fluctuation associated with the detection process, all optical detection systems would operate at the "quantum" or photon noise limit. Unfortunately, there are several other significant sources of noise in a detector circuit, the most important being the load resistor and the amplifier.

4.2 The Detector Circuit: Resistor and Amplifier Noise

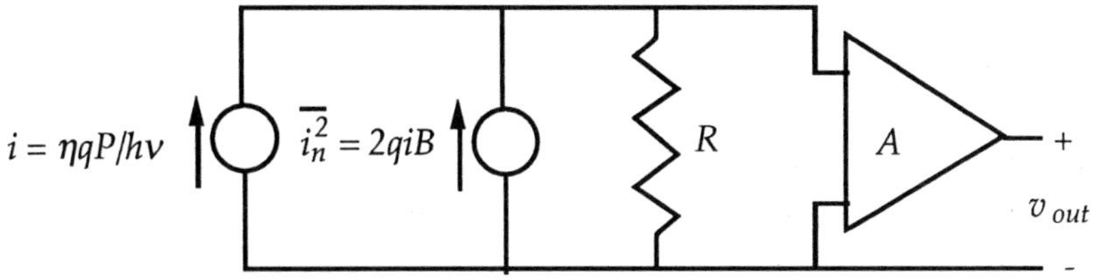

Figure 4.3 Equivalent circuit for photodetection.

To *measure* or process the information from the detector, we need a load resistance R followed by an amplifier with gain A to raise the signal to a level suitable for processing or for transmission to another location. The equivalent circuit in Figure 4.3 shows this arrangement with the expressions for the signal and shot noise currents. The resistance R includes the effective input resistance of the amplifier and must be set to a value that, combined with the ever-present detector capacitance, yields the required bandwidth B.

Just as in the case of blackbody radiation from an absorbing surface, the resistor R exhibits an electromagnetic wave emission capability. Actually it only emits power into *one* mode of *space,* that associated with a single-pair transmission line, as shown in Figure 4.4. Here the transmission line is terminated at each end with matched resistors R and the mode density in the frequency domain (see Problem 1.2) may be shown to be $dN = (2L/c)df$, which is the one-dimensional version of Eq. (1.11). Since we are concerned with the frequency of electrical circuit

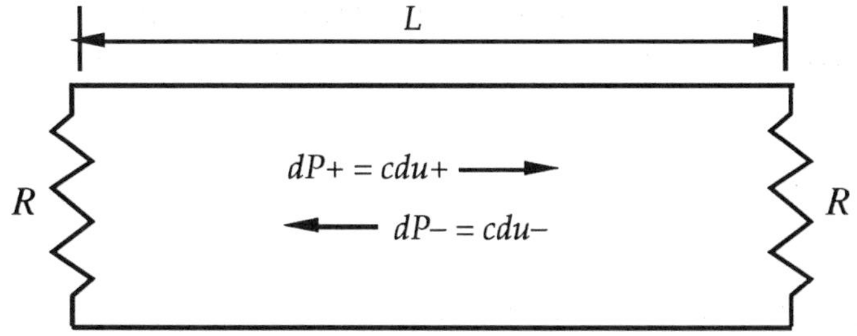

Figure 4.4 Thermal energy and power flow on transmission line.

currents, we shall again use f to distinguish it from optical or infrared frequencies. The energy per mode from Eq. (1.8) becomes kT since $hf << kT$, and the energy per unit length becomes

$$du = kTdN / L = (2kT / c)df \tag{4.7}$$

One-half of this energy is flowing to the right and the other half to the left at a velocity c, so that the respective power flows, $dP+$ and $dP-$, are each given by $cdu+ = kTdf$. If we denote B as the effective electrical bandwidth of a circuit, then the resistor must be emitting and absorbing a total thermal power of kTB. Now this power is not constant, since we know from Eq. (1.1) that the probability distribution of expected energies on the transmission line obeys the Boltzmann factor. Since the probability of observing a specific energy is Boltzmann distributed so also is the instantaneous power P. We may thus write

$$p(P) = \frac{1}{\overline{P}} e^{-P/\overline{P}} \tag{4.8}$$

and the mean square fluctuation of P, from Eq. (4.3), becomes

$$\begin{aligned}
\overline{\Delta P^2} &= \overline{P^2} - (\overline{P})^2 \\
\overline{P^2} &= \int_0^\infty P^2 p(P) dP = \int_0^\infty P^2 \frac{e^{-P/\overline{P}}}{\overline{P}} dP = (\overline{P})^2 \int_0^\infty x^2 e^{-x} dx = 2(\overline{P})^2 \\
\therefore \overline{\Delta P^2} &= (\overline{P})^2
\end{aligned} \tag{4.9}$$

Therefore the rms power fluctuation is equal to the mean or kTB. This fluctuation is the available noise power from the resistor, and it is a simple circuit exercise to show that the equivalent shunt noise current is given by

$$\overline{i_n^2} = \frac{4kTB}{R} \tag{4.10}$$

The exponential distribution of expected power corresponds to a Gaussian distribution of expected current and the thermal noise current behaves exactly the same as the shot noise current in Eq. (4.6), but with $i = 0$.

An alternative way of deriving the Johnson or thermal noise produced by a resistor uses the detailed statistics of the electron motion in the resistor material. Following the treatment of *(Yariv, 1991)*, we consider the rectangular resistor structure shown in Figure 4.5(a), which contains an electron density n and has dimensions as shown. Each electron moves randomly and has an average collision time, t_C, and

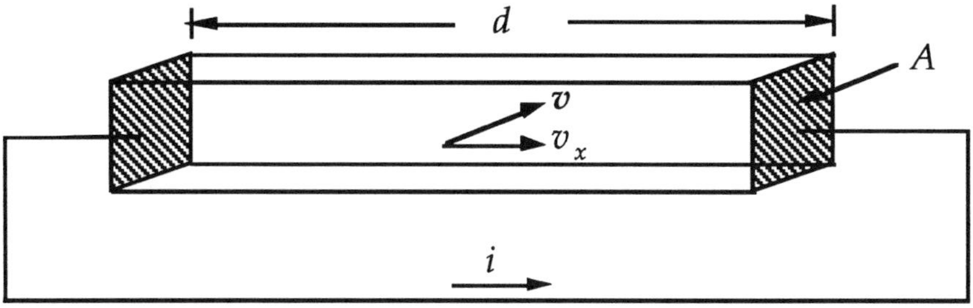

(a)

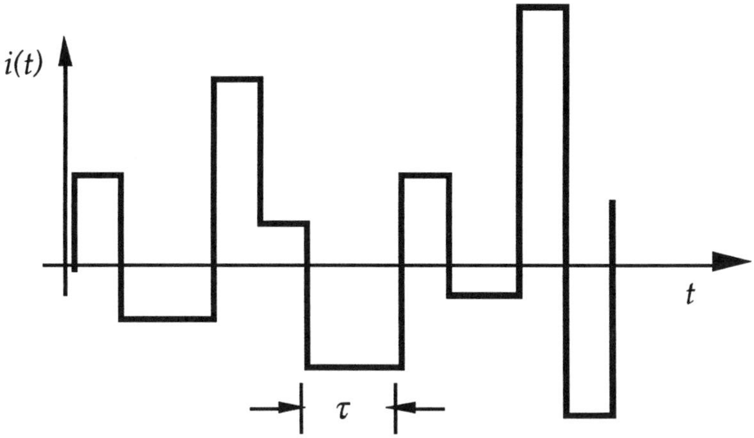

(b)

Figure 4.5. (a) Model of simple resistor showing the x-component of electron velocity. (b) Current waveform in external circuit due to single electron. τ is a particular value of the collision time which has average value, t_C.

therefore a mobility (velocity per unit electric field) given by $\mu = qt_C/m$, from standard semiconductor theory. The resistance of the element R is then given by

$$R = 1/G = 1/(\sigma A/d) = 1/[(nq\mu)A/d)] \tag{4.11}$$

which we shall relate to the noise current as follows.

As shown in Figure 4.5(a), a curren i is induced in the external short-circuit made up of the individual currents produced by each separate electron, one of whose contributions is plotted in Figure 4.5(b). Here the electron moves for a time t between collisions, inducing a square pulse of current in the output circuit of amplitude $i(t) = qv_x/d$. This current results from the time variation of the induced charge on the contact electrodes at either end of the resistor. Note that if t were the full transit time for the electron over the length of the resistor, $t = d/v_x$, then the charge transferred would be $i(t)t = q$, corresponding to a full electron charge transferred from one electrode to the other. We use the *shot noise* expressions from Eq. (4.4) and first calculate for a single electron,

$$|i(\omega)| = \left|\int_0^\tau i(t)e^{j\omega t}dt\right| = \left|\int_0^\tau \frac{qv_x}{d}e^{j\omega t}dt\right| = \frac{qv_x\tau}{d}; \quad for \quad \omega \ll \frac{1}{\tau} \tag{4.12}$$

We then average over v_x and τ, giving a mean square spectral noise current density of

$$\overline{(i_n{}^2)_f} = 2r\overline{|i(\omega)|^2} = 2r\frac{q^2\left(\overline{v_x{}^2}\right)\left(\overline{\tau^2}\right)}{d^2} \tag{4.13}$$

with $r = nAd/t_C$, the mean pulse rate, given by the total electron count, $N = nAd$, divided by the mean collision time, t_C. and assuming that v^2 and τ^2 can be averaged independently. The two mean square averages become

$$\frac{1}{2}m\overline{v_x^2} = \frac{1}{2}kT; \quad \overline{v_x^2} = kT/m$$
$$\overline{\tau^2} = 2\tau_C^2 \quad since \quad p(\tau) = \frac{1}{\tau_C}e^{-\tau/\tau_C}, \tag{4.14}$$

the latter relation from the exponential statistics of t using Eq. (4.9). Some straightforward algebra using Eq. (4.11) leads to the final relationship,

$$(\overline{i_n^2})_f = 4kT(\frac{q\tau_C}{m})\frac{qnA}{d} = 4kT\frac{\mu qnA}{d} = 4kT/R \tag{4.15}$$

which is the result we obtained for Johnson noise using thermodynamic arguments. Thus, thermal or Johnson noise may be physically described as a form of shot noise. In Chapter 5 we will examine a similar relationship for a semiconductor *p-n* junction in thermal equilibrium.

Returning to our circuit noise analysis, we add a new noise current source to Figure 4.3, which we shall represent as the noise current from a resistor R_{in} at temperature, T_N, which we call the effective noise temperature. This will allow us later to attribute the amplifier noise to the effective input resistance R_{in} as shown in Figure 4.6. Ideally we would like R_{in} to be as large as possible to yield the lowest noise current, but at the same time the net shunt capacitance of the detector input circuit limits the bandwidth to $B = 1/2\pi R_{in}C$, thus establishing a maximum value of R_{in}. We must now determine the value of the effective input noise temperature as determined by the amplifier performance.

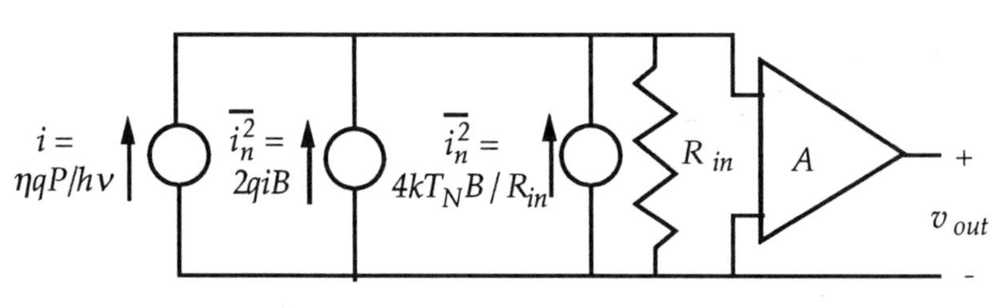

Figure 4.6 Detector circuit including resistor and amplifier noise.

4.3 The Transimpedance Amplifier

We now consider the *transimpedance* amplifier which is usually the amplifier of choice for photodetection circuits, where we are amplifying a continuous analog signal. The transimpedance amplifier has an effective input noise temperature, T_N, which can be *lower* than the physical temperature of the circuit, and with the appropriate choice of R_{in} has a uniform frequency response up to the required bandwidth. By use of feedback, the apparent input resistance of the amplifier can be made lower than the true physical resistance in the circuit, thus reducing the thermal noise current contribution. Consider a *noiseless* operational amplifier of voltage gain, A, connected as in Figure 4.7(a), showing a negative feedback resistor, R_f, with its associated noise current source. The equivalent circuit is shown in (b) with the effective input resistance, R_{in}, given by R_f/A. Since the input voltage v_{in} is $iR_{in} = iR_f / A$, the output voltage is $v_{out} = Av_{in} = iR_f$. This simple relationship explains the name "transimpedance."

If amplifier A were completely noiseless, the noise current would be that of the resistor R_f, a marked reduction from that of a true physical resistor R_{in}. Actually, the amplifier is not noiseless and for a field-effect transistor (FET) input stage, the noise currents are shown in Figure 4.8. The "pinched-off" channel of the FET in (a) produces a noise current equal to that of a resistor of value $R = 1/\gamma g_m$, with g_m being the transcon

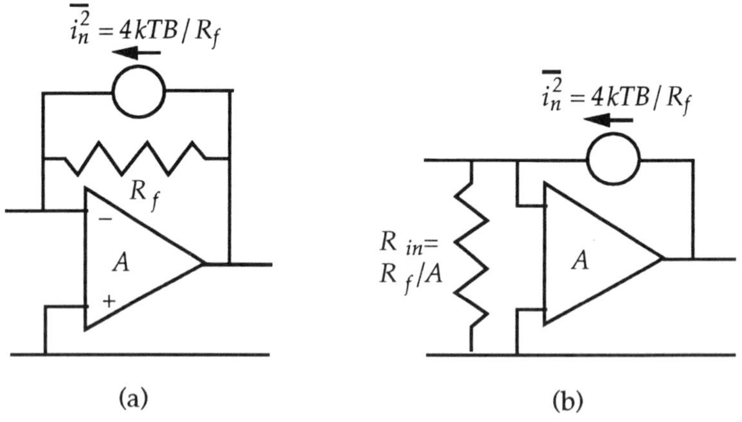

Figure 4.7 (a) Transimpedance amplifier and (b) equivalent circuit.

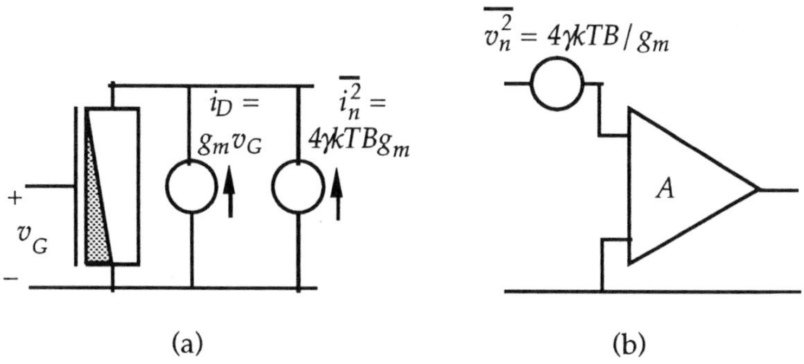

Figure 4.8 Field-effect transistor showing (a) output signal and noise currents and (b) Equivalent input noise voltage.

ductance and the factor γ ranging from 0.5 to 1.5, depending on the type of FET *(van der Ziel, 1970, pp. 72-76).* Usually the gain in the first FET stage of the amplifier is large enough that the following stages contribute negligible additional noise, so the noise voltage in Figure 4.8(b) is the appropriate input term for the amplifier noise contribution. Converting this voltage source to an equivalent current source shunting the input resistance R_{in}, the two terms become

$$\overline{i_n^2} = \frac{4kTB}{R_f} + \frac{4\gamma kTB}{g_m R_{in}^2} = \frac{4kTB}{R_f}\left[1 + \frac{\gamma}{g_m R_{in}^2}\right] = \frac{4kTB}{R_{in}}\left[\frac{1}{A} + \frac{\gamma}{g_m R_{in}}\right] \qquad (4.16)$$

If we attribute all the noise to the resistor, R_{in}, then we may write the following equation for noise temperature:

$$T_N = T\left[\frac{1}{A} + \frac{\gamma}{g_m R_{in}}\right] \qquad (4.17)$$

As an example, if $A = 10$, $R_{in} = 10{,}000\ \Omega$, and $g_m = 10^{-3}$ *mhos*, then for $T = 300$ K, T_N becomes 60 K. With this background we shall henceforth specify the detector amplifier (often called the preamplifier) performance in terms of R_{in} and T_N. Alternatively, we may use the effective input noise current as a performance measure as discussed in Section 4.6.

Although the transimpedance amplifier offers probably the best low-noise performance, the usable amplifier gain and feedback resistance are limited by amplifier stability considerations at very high frequencies as well as the effective shunt capacitance across the feedback resistor. A stray capacitance, C_S, would result in an effective input shunt capacitance across R_{in} of AC_S or *1.0 pf* if the stray capacitance is *0.1 pf* and the gain is *10*. This reduction of the feedback impedance by the gain factor, thus increasing the apparent capacitance, is known as the Miller effect, and dictates careful design trade-offs to obtain low-noise performance at the appropriate bandwidth.

The detailed noise behavior becomes more complicated near the upper frequency limits, since R_{in}^2 in Eq. (4.16) must be replaced by $|Z_{in}|^2$. Since Z_{in} contains the shunt input capacitance including the Miller effect term, the noise contribution of the second term increases rapidly near the bandwidth limit. See, for example, *(Wroblewski, 1988)*.

Although beyond the scope of this text, there is another source of noise in amplifiers at low frequencies. Sometimes called *"1/f"* or *flicker* noise, its spectral density falls off roughly inversely with frequency up to a "noise corner" frequency, where it becomes negligible compared to the thermal noise. The noise mechanism is generally attributed to carrier trapping in either surface or bulk energy states in the semiconductor. "Corner" frequencies are in the *kHz* to *MHz* region depending on the particular type of transistor.

4.4 Signal-to-Noise Voltage and Noise Equivalent Power

We are now ready to characterize the optical detector system in terms of the ratio of the signal voltage to the noise voltage at the output of the amplifier. This quantity is a crucial measure of system performance, because it tells us the relative values of the measured optical power and the apparent fluctuation in the power due to additive noise. Since we also know that the noise voltage is Gaussian distributed with zero mean, we will later be able to calculate such quantities as detection probability in terms of the known statistics. Since the amplifier in Figure 4.6 is noiseless, the signal-to-noise voltage, $(S/N)_V$, at the output is the same as the ratio of the detector signal current i_S to the rms noise current $\sqrt{\overline{i_n^2}}$, where we add the shot and amplifier mean square noise currents.

The signal current and mean square noise current then become

$$i_S = \eta q P_S / h\nu$$
$$\overline{i_n^2} = 2\eta q^2 (P_S + P_B) B / h\nu + 4kT_N B / R_{in} \tag{4.18}$$

where the total optical power has been separated into the two components P_S, the signal power or change in power which is to be measured, and P_B, the background power, which can be present in the absence of any signal. The $(S/N)_V$ is proportional to the optical signal *power*, from Eq. (4.18) and it is convenient to define a quantity, *NEP*, the noise equivalent power, which is that power which yields a $(S/N)_V = 1$.

The two most common operating modes for photodetection are the *background-limited* and *amplifier-limited* cases. In the former, the shot noise due to background radiation striking the detector is greater than that produced by the amplifier. An example is a thermal radiation measurement in the *10-μm* wavelength band. Conversely, amplifier-limited performance is generally the case at the shorter wavelengths, where background is negligible. The prime example is in direct-detection fiber optic communication. The signal-to-noise voltage becomes the ratio of the signal power to the noise-equivalent-power, P_S/NEP, and we determine the appropriate *NEP* by setting the signal current equal to the rms noise current in Eq. (4.18), yielding

$$\textit{Background-Limited:} \quad (NEP)_{BL} = \sqrt{\frac{2h\nu B P_B}{\eta}}$$
$$\textit{Amplifier-Limited:} \quad (NEP)_{AL} = \frac{h\nu}{\eta q}\sqrt{\frac{4kT_N B}{R_{in}}} \tag{4.19}$$

where we have assumed that the signal is much smaller than the background in the first case. The amplifier-limited case may be rewritten to take into account the limitation on R_{in} by the detector circuit capacitance C. Because, for a simple single-pole circuit, $R_{in} = 1/2\pi BC$, the *NEP* becomes

$$(NEP)_{AL} = 2\frac{h\nu B}{\eta q}\sqrt{2\pi k T_N C} \tag{4.20}$$

and we see the importance of minimizing the capacitance in the detector input circuit. Although we use the term "amplifier-limited," we refer to noise generated by *both* the amplifier and the physical detector load resistance.

There is a third noise-limit regime called signal- or "photon-noise"-limited performance. This occurs when the $(S/N)_V$ is limited by the shot-noise produced by the signal power itself. In this case, from Eq. (4.18), the signal-limited noise equivalent power, $(NEP)_{SL}$, becomes $2h\nu B/\eta$, but since the noise is a function of the signal power, the $(S/N)_V$ only increases as the square-root of the signal power. Usually, simple detection systems operating in this mode, such as the photomultiplier discussed in Chapter 5, use "photon-counting" or pulse counting of the output where *NEP* is not a useful concept. Later, we use $(NEP)_{SL}$ for the analysis of detection systems using avalanche photodiodes and optical (laser) pre-amplifiers, where signal-induced noise makes a significant contribution to the total noise output.

4.5 The Integrating Amplifier: Sampled Data Systems and Noise Equivalent Electron Count

Thus far we have assumed a continuous analog signal and specified a uniform frequency response determined by the product of the detector capacitance and the amplifier input resistance. The circuit of Figure 4.9 shows an amplifier with only the capacitance of the detector, amplifier, and associated wiring as the input impedance. This is called an *integrating amplifier* since the input and output voltages measure the integral of the detector current. With an appropriate differentiating filter at the output, a uniform analog signal response could be attained, but at a serious sacrifice in dynamic range, with low-frequency components easily saturating the amplifier. In contrast, if we wish to sample the total charge in individual pulses of a data stream, then the integrating amplifier is the method of choice.

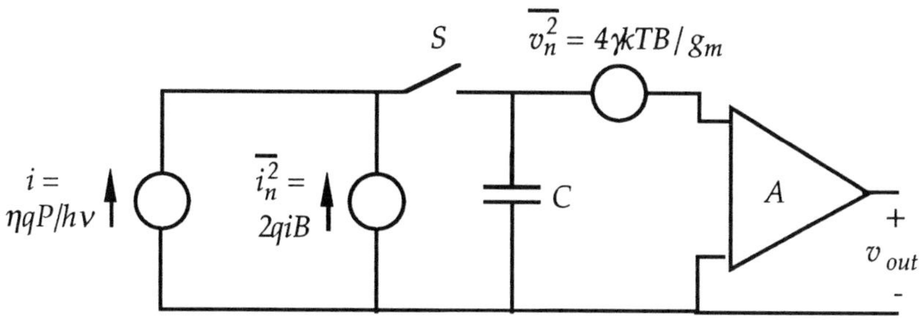

Figure 4.9 Integrating amplifier with FET input stage.

In Figure 4.9, if we sample the current for a short time τ by closing and then opening switch S, then a signal charge Q_S will charge the capacitance C, producing a step of voltage $v_S = Q_S/C = Nq/C$, where q is the electron charge and N the number of electrons. The shot noise generator, $2qiB$, will produce a fluctuation in charge equivalent to $N^{1/2}$ as detailed in our original derivation. The added noise, above this photon noise, is determined by the effective input noise of the FET amplifier. For a signal-to-noise ratio of unity we may write

$$v_S{}^2 = N^2 q^2 / C^2 = \overline{v_n{}^2} = 4\gamma kTB / g_m$$
$$\therefore \quad N = \frac{C}{q}\sqrt{\frac{2\gamma kT}{g_m \tau}} \quad with \quad B = 1/2\tau \tag{4.21}$$

using the earlier derived relation between the sampling time and the bandwidth. The quantity N is called the *noise equivalent electron count*, written variously as *NEN* or *NEE*. For a given quantum efficiency η, we can also write a *noise equivalent photon count* as $NE\Phi = NEE/\eta$. Taking typical values for a detector circuit such as $C = 1$ *pf*, $T = 300$ *K*, $g_m = 10^{-3}$ *S*, and $\tau = 10^{-6}$ *sec* yields *NEE* = approximately *17*. Thus, for example, if there were *25* electrons in a sample, there would be a shot noise fluctuation of *5* and an amplifier noise contribution of *17*. The total rms fluctuation over a series of measurements would be $NEE = (5^2+17^2)^{1/2} = 18$. For a quantum efficiency η of *0.5*, the noise equivalent photon count, $NE\Phi$, would be *36*.

Inherent in this performance calculation is the assumption that we measure the *change* in output voltage during the time the switch is closed. For repetitive sampling, if we reset the capacitor voltage to a reference value after each sample, there will always be a finite resistance in the reset circuit and the capacitor voltage will have a fluctuation or uncertainty of $\sqrt{\overline{v_n^2}} = kTC$ (see problem 4.3). This uncertainty can introduce large errors if only the final sampled voltage is measured. Sampling the voltage just before and after the charge transfer, called *correlated double sampling,* eliminates this thermal noise effect (see Sections 8.3 and 8.4).

We have previously invoked the Nyquist sampling theorem to justify the relationship between output bandwidth B and sampling or integration time τ, writing $B = 1/2\tau$. To justify this assumption, we consider a square pulse of width τ, which is the impulse response of a network that integrates over τ. The power spectral density, S_f, of such a pulse, the squared modulus of the Fourier transform, is

$$S_f = \left| \frac{1}{\tau} \int_{-\tau/2}^{\tau/2} e^{j2\pi ft} dt \right|^2 = \left| \frac{\sin \pi f\tau}{\pi f\tau} \right|^2$$

The pulse and its power spectrum are plotted in Figure 4.10 and the area under the power response curve may be calculated to be $1/2\tau$, which we shall use as the effective bandwidth of the network or alternatively as the

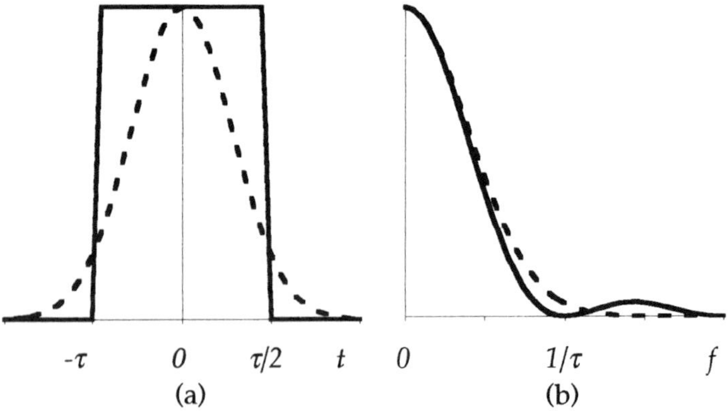

Figure 4.10 (a) Normalized current pulse shape and (b) power spectral density. Solid line, square pulse; dashed line, Gaussian pulse

network bandwidth required for effective detection of a pulse of width τ. Also shown for comparison is a Gaussian pulse, $exp(-2\pi t^2/\tau^2)$, with its associated power spectrum, $exp(-\pi f^2\tau^2)$ which also has area or first moment, $1/2\tau$. Thus, $B = 1/2\tau$ is a reliable approximation for the required network bandwidth for a typical current pulse.

4.6 Effective Input Noise Current, Responsivity, and Detectivity

To this point we have characterized the preamplifier in terms of an effective input resistance and noise temperature. From a system design point of view, commercially available amplifiers are usually specified in terms of the normalized or "spot" noise current measured in $A/\sqrt{Hz}$, usually quoted in $pA/\sqrt{Hz}$, with $1\ pA = 10^{-12}\ A$. The effective input noise current then becomes

$$(i_n)_{eff} = \sqrt{\overline{i_n^2}} = \sqrt{\frac{4kT_N B}{R_{in}}} = \left\{\sqrt{\frac{4kT_N}{R_{in}}}\right\}\sqrt{B} = (i_n)_{\sqrt{f}}\sqrt{B} \tag{4.22}$$

and the quantity, $(i_n)_{\sqrt{f}}$ can be found from Figure 4.11.

Example. The transimpedance amplifier example at the end of Section 4.3 used an input resistance of $R_{in} = 10{,}000\Omega$, and had an effective input noise temperature of $T_N = 60\ K$. From Figure 4.11, the effective input noise current density is $0.6\ pA/\sqrt{Hz}$. Assuming a $1\ MHz$ bandwidth, compatible with TV imaging rates, the net input noise current is then $600\ pA$ or approximately 4×10^9 *electrons/sec*. The sampling time corresponding to this bandwidth is $0.5\ \mu sec$, and thus the noise or rms count fluctuation in any sample is *2000!* Thus, even with the reduced effective temperature, the performance is far removed from "photon" counting.

Another descriptor for detectors is the *responsivity* $\Re$, the current output per unit optical power given by $\eta q/h\nu = \eta\lambda\ (\mu m)/1.24$ in A/W. The amplifier noise limited *NEP* then becomes

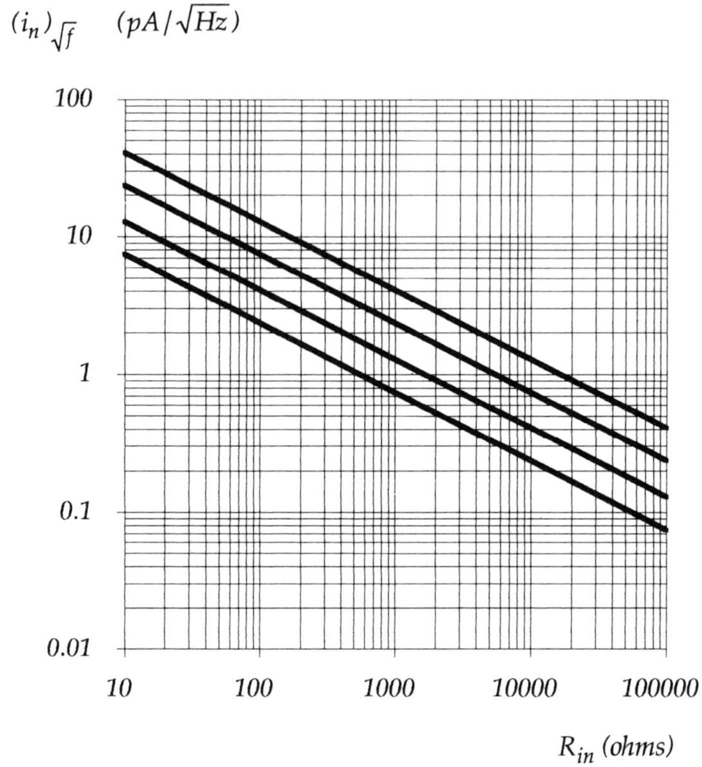

Figure 4.11 Normalized noise current $(i_n)_{\sqrt{f}}$, *vs* R_{in} for $T_N = 10, 30, 100,$ *and* 300 *K.* $1\ pA = 10^{-12}$ *amp.*

$$(NEP)_{AL} = (i_n)_{eff} / \Re = (i_n)_{\sqrt{f}} \sqrt{B} / \Re \tag{4.23}$$

Finally we define *detectivity* D as the reciprocal of the *NEP*, with dimension W^{-1}. This figure of merit increases when the sensitivity or ability to *detect* signals increases Many detectors are also specified in the literature in terms of their *specific detectivity* D^* *(pronounced D-star)*. This quantity is D normalized to a standard detector area A of $1\ cm^2$, and bandwidth, B, of $1\ Hz$. We may then write D^* for the background-limited and amplifier-limited cases as

$$D^* = D\sqrt{AB} = \frac{\sqrt{AB}}{NEP} \quad W^{-1} cmHz^{1/2}$$

$$(D^*)_{BL} = \sqrt{\frac{\eta}{2h\nu BP_B}}\sqrt{AB} = \sqrt{\frac{\eta}{2h\nu(P_B / A)}} \tag{4.24}$$

$$(D^*)_{AL} = \frac{\eta q}{h\nu}\sqrt{\frac{R_{in}}{4kT_N B}}\sqrt{AB} = \Re\sqrt{\frac{R_{in}A}{4kT_N}}$$

using Eq. (4.19). In the background-limited case we see that D^* is controlled by the background power density striking the detector. There is thus a family of limiting curves for D^* in the long-wavelength region where the background is typically at *300 K* and the background intensity is a function of the optical system *f/#*, or the numerical aperture, *NA* (see *Kingston, 1978*). The amplifier-limited expression is particularly applicable to photovoltaic detectors discussed in Chapter 5, where the so-called *RA product* becomes a figure of merit.

4.7 General Signal-to-Noise Expression for Combined Signal, Background, and Amplifier Noise

We have thus far treated detection systems as operating in one of three distinct regimes: signal or photon noise limited, background noise limited, or amplifier noise limited. In some important cases, however, the signal-induced noise can become comparable or greater than the amplifier noise at high values of $(S/N)_V$. Under these conditions, it is no longer correct to set the $(S/N)_V$ equal to P_S/NEP, although the *NEP* is still a qualitative figure of merit. This is an especially important consideration in communications and radar systems, where high $(S/N)_V$ is required for extremely low error rates and high detection probabilities. In addition, avalanche photodiodes and optical amplifiers usually have a strong signal-dependent noise output component.

For the most general expression for signal to noise we take into account all noise current contributions and write

$$\left(\frac{S}{N}\right)_V = \frac{i_S}{\sqrt{\overline{i_n^2}}} = \frac{i_S}{\sqrt{2qiB + (4kT_N B / R_{in})}} \tag{4.25}$$

and using $i = \eta q(P_S+P_B)/h\nu$, we obtain, after some manipulation,

$$\left(\frac{S}{N}\right)_V = \frac{P_S}{\sqrt{\frac{2P_S h\nu B}{\eta} + \frac{2P_B h\nu B}{\eta} + \left(\frac{h\nu}{\eta q}\right)^2 \frac{4kT_N B}{R_{in}}}} \tag{4.26}$$

This result may now be written in terms of our previous definitions of $(NEP)_{BL}$ and $(NEP)_{AL}$, and the quantity $(NEP)_{SL} = 2h\nu B/\eta$, yielding

$$\left(\frac{S}{N}\right)_V = \frac{P_S}{\sqrt{P_S \times (NEP)_{SL} + (NEP)_{BL}^2 + (NEP)_{AL}^2}} \tag{4.27}$$

We use this expression in Chapter 7.

Problems

4.1 In Problems 1.9, 1.10, and 1.11 on thermography, the detector received approximately 1.5×10^{-5} W in the 10-μm band, and the fractional change in power with temperature was found to be 0.016 K^{-1}. Denoting the change in power as P_S and using a 1-MHz bandwidth, suitable for a TV display, find:

(a) The mean square noise current due to the background for a unit quantum efficiency photon detector.
(b) The signal current, i_S, due to a temperature change, ΔT.
(c) The noise equivalent temperature change, $NE\Delta T$.

4.2 The detector in the previous problem has an output capacitance of 10 pf and a load resistor is chosen so that the half-power response is at 1 MHz. Find the mean square noise current due to the resistor. Find the new $NE\Delta T$.

4.3 Another way to show the spectral density of Johnson noise is to consider a capacitor C in parallel with a resistance R. The stored thermal energy of the capacitor is potential and has the value $kT/2$. This energy results in a mean square noise voltage at the capacitor terminals determined by the value of C. If $(v_n^2)_f$ is the spectral density of the mean square noise voltage, show that its value is given by $4kTR$ by integrating it over the power response of the RC circuit yielding the total mean square fluctuation, *i.e.*

$$\overline{v_n^2} = \int_0^\infty (v_n^2)_f \, |H(f)|^2 \, df$$

4.4 The responsivity, $\Re$, of a detector is defined as the number of amps per watt of incidental optical power. Write a numerical expression for $\Re$ in terms of the quantum efficiency η and the wavelength λ in μm.

4.5 A unit quantum efficiency ideal detector at a *10-μm* wavelength is operated at *77 K* with an integral bias resistor of *1000 Ω* in the same temperature bath. The detector is coupled by a cable to an external transimpedance amplifier with an input resistance, *300 ohms*, and effective input noise current of *3 picoamp/√Hz*. Find:

(a) R_{in} and T_N as seen by the detector.
(b) The overall effective input noise current in *A/√Hz*.
(c) The $(NEP)_{AL}$ for a bandwidth of *1 MHz*.

4.6 Write an expression for the number of received photons/pulse required for a $(S/N)_V = 1$, for a pulse of duration τ with bandwidth $B = 1/2\tau$, given the effective input noise current $(i_n)_{\sqrt{f}}$.

Chapter 5
Real Detectors:
Vacuum Photodiodes and Photomultipliers, Photoconductors, Junction Photodiodes, and Avalanche Photodiodes

The ideal photon detector we have discussed is an excellent model for many real detectors, except that we must consider in detail the expected capacitance, the frequency response, and the spectral behavior of the quantum efficiency. We first consider the vacuum photodiode and photomultiplier, the latter having adequate internal gain or electron multiplication such that amplifier noise is not a limiting factor. Unfortunately, the photoelectric effect used in such devices is only efficient at near-visible and shorter wavelengths and, beyond about a *1-μm* wavelength, internal electron excitation in a semiconductor is the only available alternative. This class of detectors includes photoconductors, basically resistances whose values change with illumination, and semiconductor photodiodes, where the reverse or saturation current is controlled by the incident radiation. Photodiodes may also be operated in the "photovoltaic" mode where the open-circuit voltage is a measure of the incident power, a small-signal version of the solar cell. Again, a device such as a semiconductor diode is close to the ideal photon detector but still suffers from the limitations of amplifier noise. We conclude the chapter with a discussion of the "avalanche" photodiode, which supplies internal electron gain, thus offering in some cases marked improvement over the simple diode.

Although we are discussing real detectors with performances approaching ideal "photon" counting, the reader is reminded that there are a family of simple though far less sensitive devices known as "thermal" detectors. These include bolometers, thermocouples, and pyroelectrics, which exhibit a resistance or voltage output which is a function of the temperature of a small sensing element that is in turn heated by the incident optical or infrared radiation.*(Kingston, 1978).* Since they only require energy absorption at the desired wavelength, their spectral response is effectively unlimited. Since they depend upon heating and cooling for their operation, the frequency response is limited.

5.1 Vacuum Photodiodes and Photomultipliers

The vacuum photodiode, which uses the "external" photoelectric effect first analyzed by Einstein, consists of a "photocathode" and an electron collector or "anode." For simplicity we shall use the plane-parallel configuration of Figure 5.1, although practical devices often consist of a half-cylinder photocathode with a thin wire along the cylinder axis as the anode. The quantum efficiency falls sharply for wavelengths beyond about $0.7\ \mu m$ and is negligible beyond about $1.1\ \mu m$. Figure 5.2 is an approximate representation of the maximum quantum efficiencies available with different photocathode materials. Values for the long-wavelength region beyond about $0.9\ \mu m$ are subject to the exposure history of the photocathode. Exposure to visible light can permanently degrade the performance of many of the long-wavelength coatings.

Although the vacuum photodiode approaches the ideal detector in its behavior, it has a frequency response limited by the electron transit time and the cathode/anode capacitance. In addition, for long- wavelength photocathodes, there can be appreciable "dark" current, electrons emitted by thermal energy rather than absorbed optical energy.

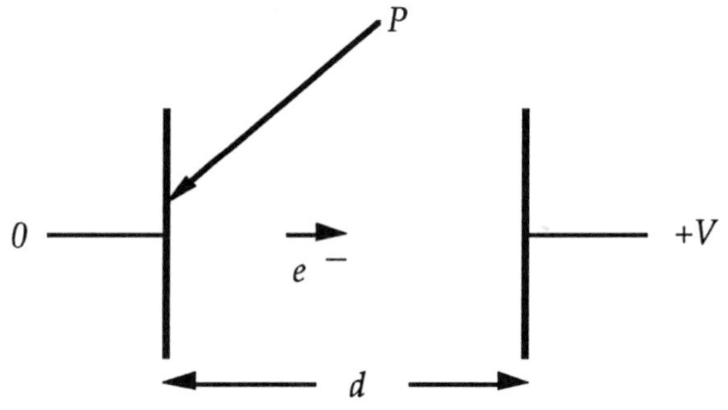

Figure 5.1 Simple representation of vacuum photodiode.

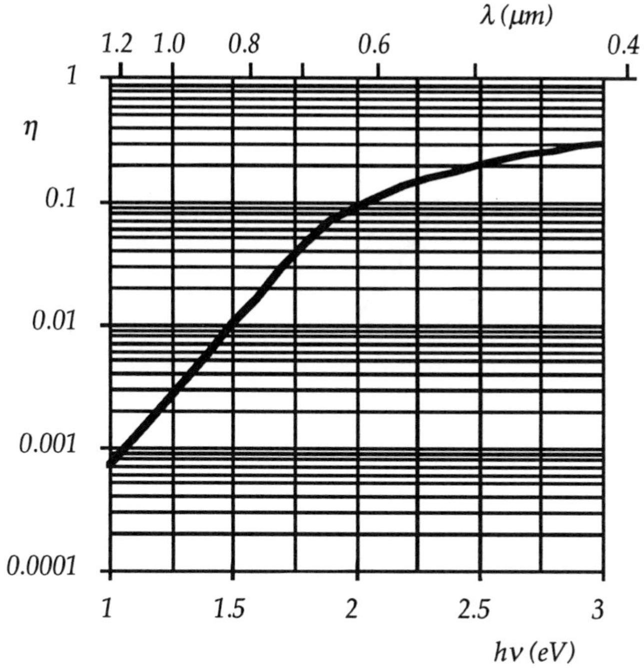

Figure 5.2 Typical maximum quantum efficiencies for a selection of commercial photocathodes.

The frequency response is determined by the shape of the electrical output pulse for an impulse of optical power input. Such an impulse produces a charge packet, which moves with constant acceleration from the cathode to the anode. The current flow in the external circuit is proportional to the electron velocity, which increases with time as $v = at$, where a is given by $F/m = qE/m = qV/md$. The resultant triangular current pulse is given by $i(t) = 2qt/\tau^2$, from $t = 0$ to $t = \tau$, where τ, the transit time, is given by $\tau = d(2m/qV)^{1/2}$. The power response in the frequency domain is obtained from the square of the Fourier transform,

$$|\,i(\omega)\,|^2 = \left| \int_0^{\tau} \frac{2qt}{\tau^2} e^{j\omega t} dt \right|^2 = q^2 \frac{4}{(\omega\tau)^4} [4\sin^2(\omega\tau/2) + (\omega\tau)^2 - 2\omega\tau\sin\omega\tau] \tag{5.1}$$

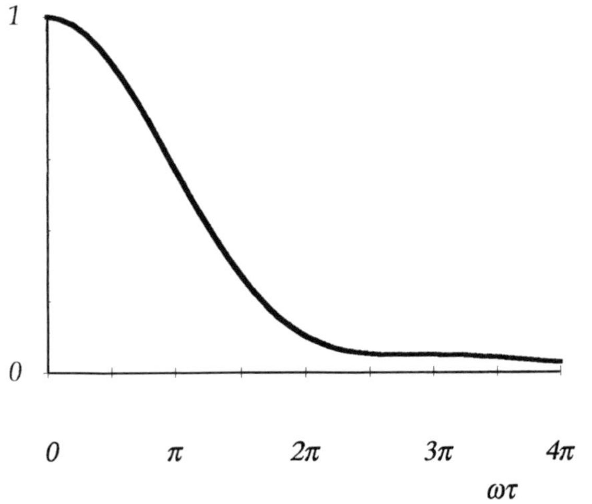

Figure 5.3 Vacuum photodiode response as a function of $\omega\tau$.

which is plotted in normalized form in Figure 5.3 as a function of $\omega\tau$. The response is seen to have a half-power point very near the value $\omega t = \pi$. The resultant half-power frequency or frequency "cut-off," f_{co}, then becomes $(1/2\tau)$. In a similar fashion, from Eq. 4.4, the shot noise power spectrum obeys the same form as Eq. (5.1). As an example of a typical diode, we take the parameters $d = 1\ cm$, electrode area $A = 1\ cm^2$, and $V = 300V$. The resulting frequency cut-off is $250\ MHz$ for a transit time of $2 \times 10^{-9}\ sec$ and a capacitance of $0.09\ pF$, using $\varepsilon_o = 8.85 \times 10^{-14}\ F/cm$ and the electron mass $9.1 \times 10^{-31}\ kgm$. To match the transit-time-limited frequency response to the RC response would require a resistive load of $7000\ \Omega$.

Finally we can estimate the dark current from the Richardson-Dushman thermionic emission equation, which describes the current caused by highly energetic thermal electrons that can escape over the potential barrier, ϕ, the work function of the metal. The dark current density is given by

$$J_D = 120T^2 e^{-q\phi/kT} \quad A/cm^2; \quad T \text{ in } K \tag{5.2}$$

If the photocathode has a cut-off wavelength of $0.8\ \mu m$, $\phi = 1.55\ eV$, and

J_D at room temperature becomes approximately $10^{-19}\ A/cm^2$, or about one electron per second per square centimeter. Photocathodes with longer wavelength cut-offs can have appreciable dark current as a result of the smaller ϕ. Consequently some applications, especially photon-counting using a photomultiplier as described later, require cooling of the photocathode.

As we learned in Chapter 4, the ideal photon detector is generally amplifier noise limited in the visible, where the thermal background is low. As a result, to obtain photon-counting or signal-noise-limited operation, we must somehow obtain noiseless gain before electrical amplification. This is accomplished in the structure shown schematically in Figure 5.4, the photomultiplier. A single photoelectron leaving the cathode is converted to an output pulse of total charge Gq, where the multiplier gain is $G = \delta^N$, where δ is the multiplication factor of the "dynode" stages, *1* through *8*. Each of these electrodes is biased approximately *100 V* positive with respect to the preceding stage and multiplication occurs because of secondary emission. The current gain in such a device can be as high as 10^6 and as a result the output pulse due to a single photoelectron can be observed. The frequency response is limited by transit time *spread* as an individual charge pulse travels through the chain, with a final contribution due to the finite pulse width caused by the charge transit from electrode number *8* to the collector, *A*. The output capacitance is about the same as that of a single diode and the dark current is of course the value obtained from Eq. 5.2 but multiplied by the current gain *G*. The shot-noise becomes

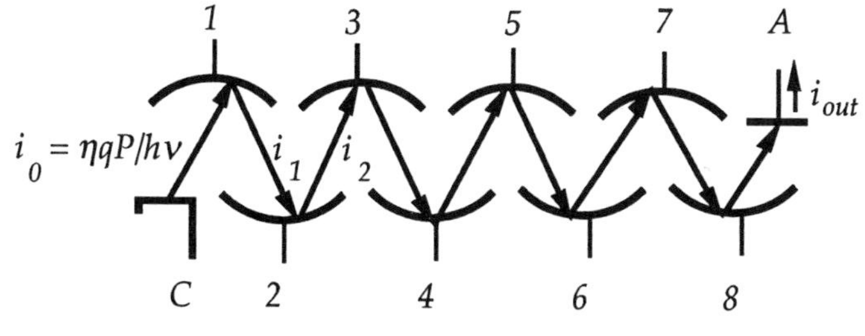

Figure 5.4 Photomultiplier. *C*, photocathode; *1 - 8*, dynodes; *A*, collector.

$$< i_n^2 >= 2q\Gamma G^2(i_S + i_D)B \qquad (5.3)$$

with i_S as the signal current and i_D as the dark current. The G^2 results from the multiplication of the photocathode current fluctuations and the Γ from the statistical nature of the dynode multiplication process. For a Poisson distributed multiplication factor with mean value δ, the value of Γ is found to be $\delta/(\delta - 1)$ *(Kingston, 1978)*. The excess noise at the output may be understood by considering the noise contributions of the successive electrodes. The photocathode, because of shot noise, contributes a term $2qG^2(i_S + i_D)B$, since the mean square noise current is multiplied by the square of the current gain G. The first dynode has an average multiplication factor δ, which if Poisson distributed, has a mean-square fluctuation of δ. Just as in the shot-noise case the added mean square noise current from the dynode becomes $2q[\delta(i_S+i_D)]B$, corresponding to a mean current of $\delta(i_S+i_D)$. But this mean square current is only amplified by $(G/\delta)^2$ rather than G^2 since it traverses one less dynode. Considering successive dynode stages, we may obtain the final result, $2qG^2(i_S+i_D)B\ [\ 1 + (1/\delta) + (1/\delta)^2 + \cdots\ ]$, and this series is easily shown to be the quantity $\delta/(\delta - 1)$, as given earlier. Typical dynode gains are the order of five, yielding an excess noise factor Γ of 1.25.

Example. We wish to count individual photoevents from a photomultiplier with an output pulsewidth of $2\ ns$. Using the same parameters as in the earlier vacuum photodiode example, we first calculate the expected rms noise current using an amplifier input resistance of $7000\ \Omega$ and a noise temperature $T_N = 300\ K$. From Fig. 4.11, the *root spectral density* is approximately $1.5\ pA/\sqrt{Hz}$, which for the required bandwidth, $B = 1/2\tau$, of $250\ MHz$ results in an rms effective input noise current of $24\ nA$. The expected single photoelectron pulse amplitude is Gq/τ, with $G = \delta^N = 5^8 = 4 \times 10^5$. The resultant current pulse is then $32\ \mu A$ or a peak pulse/rms noise current ratio of 1300. We may thus set a current threshold level well above the noise but below the expected pulse height and perform reliable "photon" counting, albeit limited by the finite quantum efficiency and any dark current contributions.

We shall consider quantitatively in Chapter 7 the relation between the threshold setting and the probability of an erroneous count due to a noise pulse for Gaussian noise. A threshold current set at *12* times the rms noise current, for example, results in a false count probability of 10^{-9} for each pulse interval or an error rate of $10^{-9}/2 \times 10^{-9}$ or one false count every *2 sec*. If the expected pulse height were a constant, then each electron emitted by the photocathode would be detected. Actually the gain is a random process as we have discussed in deriving the noise factor, Γ. In the extreme, we know that there is a finite probability that *no* electron will be emitted from the first dynode. If the statistics are indeed Poisson, then from Eq. (4.2), a δ of *5* yields a probability of nonemission of $p(0, 5) = e^{-5} \approx 10^{-2}$. Using an analysis similar to that used earlier for Γ, the fractional mean square gain fluctuation is found to be $(1/\delta + 1/\delta^2 + ...) = (\Gamma - 1)$. For $\delta = 5$, the mean square fractional fluctuation is *0.25*, and the rms fluctuation, *0.5*. The gain probability distribution is not Gaussian but we conclude that the threshold must be set well below *0.5* of the mean pulse height to assure the detection of a high fraction of the individual photoelectron events.

5.2 Semiconductors: Photoconductors and Photodiodes

We consider primarily the semiconductor junction photodiode in our systems analyses, but we first review briefly the photoconductor, which has more specialized uses although it was historically one of the first solid-state optical detectors. The photoconductor, as its name implies is a conducting element whose conductance is controlled by incident infrared or visible radiation. As shown in Figure 5.5, light striking a homogeneous semiconductor produces holes or electrons or hole-electron pairs which cause current to flow in the presence of voltage, V. There are two types of photoconductor, *intrinsic* and *extrinsic*, the former depending on across-the-gap transitions or pair production and the latter on excitation of carriers from an impurity level in the forbidden region. In either event, there are two parameters of interest, the *recombination* time τ_r and the transit time τ_t, the time for the carrier to traverse the distance ℓ in the presence of the electric field V/ℓ. The photocurrent in the presence of optical power P, becomes

$$i = \eta\left(\frac{qv\tau_r}{\ell}\right)r = \eta q\left(\frac{\tau_r}{\tau_t}\right)\frac{P}{h\nu} = \frac{\eta q G_p P}{h\nu}; \quad \eta = (1-\rho)[1-e^{-\alpha d}] \tag{5.4}$$

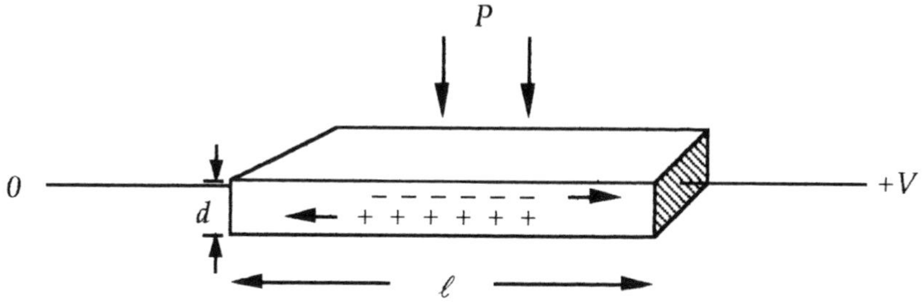

Figure 5.5 Semiconductor photoconductor.

since, as in section 4.2, qv/ℓ is the induced current flow at the electrodes during recombination time τ_r. The quantum efficiency, η, is determined by the reflectivity, ρ, at the input surface and the bulk absorption coefficient, α (see problem 5.3). The ratio of the time constants is called the photoconductive gain G_p. Similar to our treatment of the colliding electrons in section 4.2, the fractional electron charge contribution in the first parentheses is given by the ratio of the distance traveled before recombination to the length of the device. The expression may be seen to be consistent if, for example, the transit time and recombination times are equal. Then, on the average, an excited carrier moves the length of the photoconductor before recombining and contributes one electron charge to the external circuit. It is apparent that the gain can also be greater than unity, but this improved performance is accompanied by a reduced frequency cut-off, which is the inverse of the recombination time. The induced photocurrent has a mean square fluctuation called *g-r* noise for *generation-recombination,* which is given by *(Kingston, 1978, Ch. 6)*

$$<i_n^2>= 4(qG_p)iB = 4qG_p^2\frac{\eta qP}{h\nu}B \tag{5.5}$$

This may be simply interpreted as shot noise due to pulses of charge (qG_p), an extra factor of two arising from the fluctuation in pulse width due to the randomness of the recombination process. Intrinsic photoconductors are made from a myriad of materials such as PbS, PbSe, PbTe, CdS, etc., as well as HgCdTe and the well-known III-Vs, GaAs, etc., and, of course, Si and Ge. Extrinsic photoconductors are usually made by doping Si or Ge with impurities such as As or Cu and these devices, operated at cryogenic temperatures, are mostly used for long-wavelength detectors at wavelengths of $10\ \mu m$ or greater. Although appreciable gain may be obtained, this is usually with long-lifetime material and the response times are too long for many applications. The extra factor of two in the photoconductor noise output also favors junction photodiodes, which we now discuss.

The junction diode is almost identical to the vacuum photodiode in general behavior, except that it involves two types of carriers, the hole as well as the electron. Although simple step or graded junction diodes are good photodetectors, the best performance is obtained from the *p-i-n* structure shown in Figure 5.6. By keeping the doped regions thin, most of the hole-electron pairs are produced in the depletion region and carriers are not required to diffuse from the bulk, which would cause poor frequency response. The *p-i-n* structure allows a thick depletion region since there is no fixed space-charge and thus the photon interaction length is maximized. For optimum performance the light should be incident from the left, since then the dominant current through the drift region will be electrons moving to the *n*-type electrode and the higher electron drift velocity will yield higher frequency response. The most commonly used materials are Si and GaAs. The latter is generally superior except for the shorter wavelength cut-off at $0.9\ \mu m$. Silicon, in contrast, operates out to $1.3\ \mu m$, but the absorption coefficient is much smal-

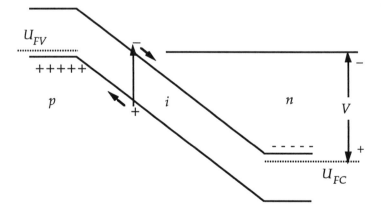

Figure 5.6 Energy band diagram for reverse-biased *p-i-n* photodiode.

Table 5.1 Important constants for Si and GaAs and the resultant performance for an *0.1- × 0.1 --mm* photodiode with a thickness $d = 1/\alpha$ at $\lambda = 0.8\ \mu m$. $\eta = (1-\epsilon^{-1}) = 0.63$

	Si	GaAs
$\alpha\ (cm^{-1})$	10^3	10^4
v_{max} (cm/sec)	3×10^6	10^7
d (for $\eta = 63\%$)	$10\ \mu m$	$1\ \mu m$
f_{co} (GHz)	1.5	50
C (pf)	0.1	1.0

ler and thus a thicker depletion layer is needed for reasonable quantum efficiency. The ternary compound, InGaAs, and the quaternary, GaInAsP, are excellent detectors at longer wavelengths, although fabrication is more complicated. Some important parameters for Si and GaAs are listed in Table 5.1, assuming an anti-reflective coating, $\rho = 0$. In some of these materials, such as GaAs, a Schottky barrier diode is used rather than a *p-i-n* structure *(Wang, 1989, p. 795 et seq.)*. In this case the *p*-type layer is replaced by a transparent metallic material such as gold or platinum.

The velocity, v_{max}, is the saturation velocity for electrons and occurs at fields of the order of $10^3\ V/cm$, or at most several volts bias. Since the velocity is *constant* across the space-charge region, the characteristic current pulse is square rather than triangular as in the vacuum diode. The resultant frequency response is then of the form $(\sin x/x)^2$, with $x = (\omega\tau/2)$. The normalized response function is plotted in Figure 5.7 and again the half-power frequency response is at approximately $1/2\tau$ as in a vacuum photodiode. The capacitance is obtained using $\kappa = 12$, the dielectric constant for both materials. There is obviously a trade-off between frequency cut-off and capacitance by varying thickness d. In fact, *50-GHz* performance in GaAs requires a much smaller sensitive area because of the capacitance limitation. As for noise, both junction and Schottky photodiodes produce simple shot noise, and the dark current is the thermally generated reverse saturation current of the device. Since these diodes are usually amplifier noise limited dark current is generally not a serious problem.

We close this section with an alternative method of operation of a a photodiode, the "photovoltaic" or open-circuit nonbiased mode. Here

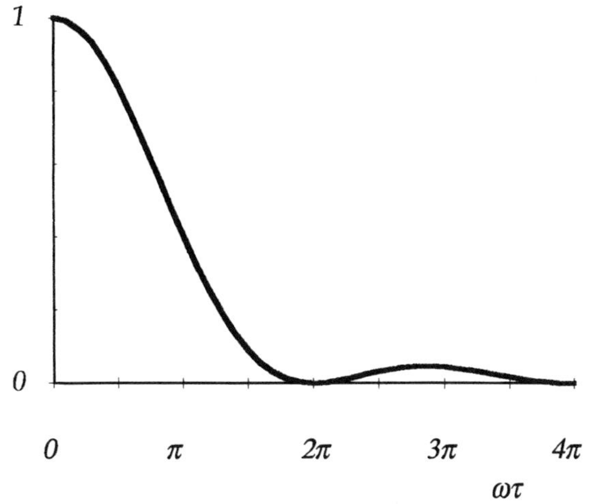

Figure 5.7 Response for junction photodiode as a function of $\omega\tau$.

we measure the voltage induced by photocreated carriers in the junction under the restriction that there is no current flow. As shown in Figure 5.8, electrons and holes produced in or near the space-charge region produce a net current flow to the left or from the n-type to the p-type material. But this current has no place to go since the diode is open-circuited. As a result, the buildup of positive charge in the p-type material and negative charge in the n-type material produces a *forward* bias and an external open-circuit voltage develops which produces a forward current equal and opposite to the photoinduced current flow. This

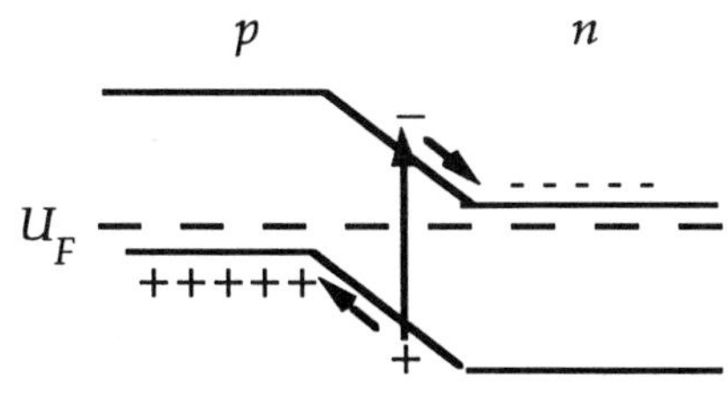

Figure 5.8 Photoinduced current flow in unbiased photodiode.

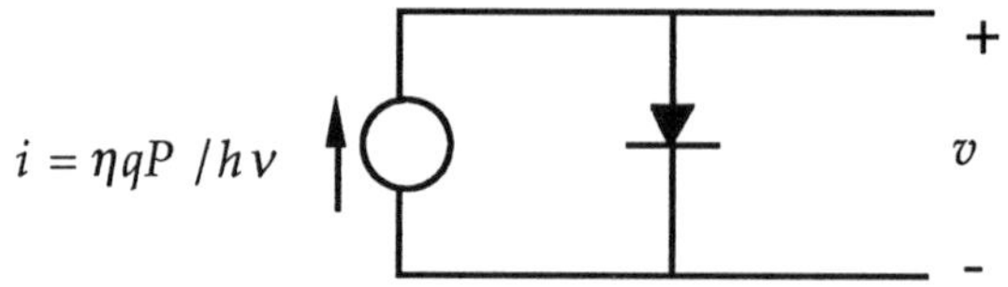

(a)

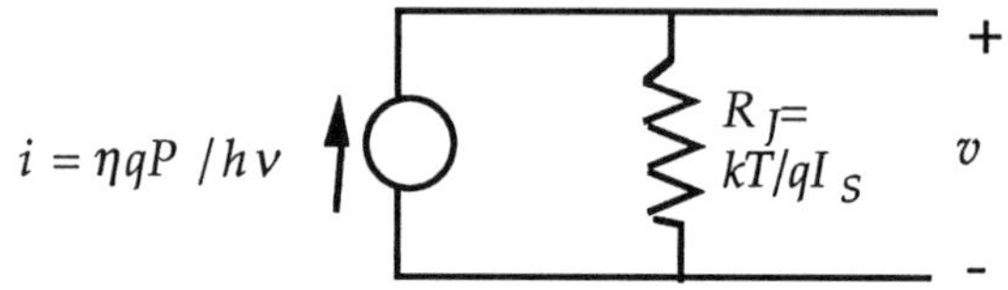

(b)

Figure 5.9 Equivalent circuit for the photovoltaic mode: (a) large signal and (b) small signal.

behavior leads to the equivalent circuit of Figure 5.9(a) which is an ideal diode current source driving the diode in the forward direction. If the reverse saturation current of the diode is I_s, then we may write

$$i = I_s(e^{qv/kT} - 1) \qquad \therefore v = \frac{kT}{q} ln\left(\frac{i + I_s}{I_s}\right) \Rightarrow \frac{kT}{qI_s} i \quad for\ i << I_s \tag{5.6}$$

The voltage response is obviously nonlinear unless the photocurrent is small compared with the saturation current, concomitant with v being much less than kT/q. For this small-signal operating mode, the final expression is seen to be the product of the junction resistance at zero bias, kT/qI_s, and the induced photocurrent. Thus the diode in the circuit may be replaced by a resistor equal to the junction resistance, as shown in Figure 5.9(b). The current fluctuation may be treated in two different ways. We can add up the shot noise due to the forward current and the reverse photocurrent, which are each equal to the saturation current in magnitude for $i << I_s$. The fluctuations are independent so we take the sum of the squares and obtain

$$< i_n^2 >= 2q(I_s + I_s)B = 4qI_sB \tag{5.7}$$

or, alternatively, since the junction is effectively in thermal equilibrium, we may calculate the Johnson noise of the junction resistance, kT/qI_s, which yields

$$< i_n^2 >= \frac{4kTB}{R} = \frac{4kTB}{kT/qI_s} = 4qI_sB \tag{5.8}$$

not surprisingly the same answer. We see then that at thermal equilibrium, thermal noise can be derived from shot noise considerations. The equivalent circuit for the photovoltaic case is thus a simple current source in parallel with a load resistor equal to the zero-bias junction resistance and having a simple Johnson noise current with the temperature T equal to the actual junction temperature. The advantages of photovoltaic operation are the simplicity of a nonbiased circuit and, for many long-wavelength photodiodes, the elimination of excess noise due to leakage currents caused by a finite bias. Since the effective input resistance is limited by the device itself, the sensitivity is determined by the zero-bias junction resistance, which becomes R_{in}, and junction temperature, which becomes T_N. Generally the following preamplifier contributes negligible noise, although we still refer to the *NEP* as amplifier-limited even if the junction noise is dominant. On the negative side, the capacitance is much higher than a biased diode and the transit time is longer because of the small depletion layer field. Many long-wavelength detectors, generally cryogenically cooled, are operated in this mode and a figure of merit often used is the *RA product* since for a given sensitive area, the larger R is, the larger the signal voltage or, conversely, the smaller the competing Johnson noise current. The significance of the *RA product* is apparent from Eq. (4.24) which relates this figure of merit to the specific detectivity D^*.

5.3 Avalanche Photodiodes

As we have seen, the semiconductor photodiode behaves like the "ideal" detector, unfortunately, it does so to the extent that amplifier noise seriously limits its sensitivity. But there is a multiplication process available in junction diodes which is similar to that of the photomultiplier, although we will discover it has serious limitations. The process is known as *avalanche multiplication,* a term that originated in gaseous electronic tubes many years ago. If the electrons or holes transiting the space-charge region gain enough energy from the electric field they can produce additional hole-electron pairs just as in the photoexcitation process. The added carriers in turn can produce further carriers and the process may be likened to boulders descending a mountain and multiplying the cascade as they descend; thus, an avalanche.

The impact ionization process, as it is called, may be described mathematically by

$$\frac{\partial I_n}{\partial x} = \alpha_n I_n(x) + \alpha_p I_p(x) = -\frac{\partial I_p}{\partial x} \tag{5.9}$$

where the junction and the current convention are shown in Figure 5.10. It helps conceptually to associate the values of the currents in this expression to the size of the corresponding arrows in the figure. Thus the electron current increases to the right as more electrons are produced by the ionization process. Conversely, the hole current *decreases* to the right, since the hole concentration increases as extra ionized holes are swept to the left. The ionization coefficients α_n and α_p are expressed in units of cm^{-1} and increase rapidly with the field E above field values of the order of $10^5\ V/cm$. In our treatment we shall assume a constant field and thus constant α_p and α_n A less restrictive treatment *(Kingston, 1978)* assumes a field variation and a consequent x-dependence of each

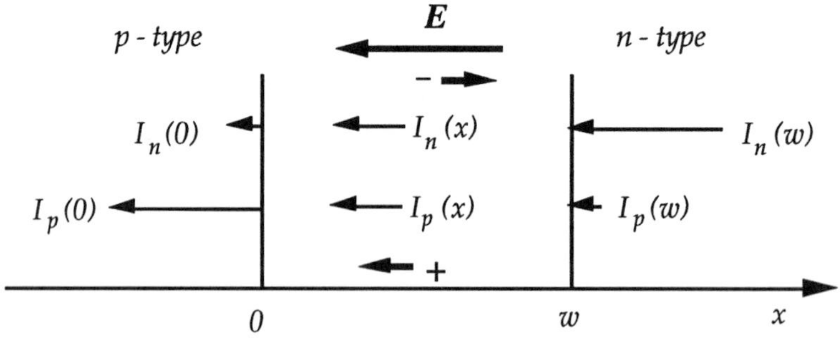

Figure 5.10 Current flow in an avalanche photodiode.

α, but yields the same results that follow for $\alpha_p = 0$ and also for $\alpha_p = \alpha_n$. The original and most general derivation is that of *(McIntyre, 1966)*.

We first consider the simplest case of operation, that in which one of the ionization coefficients is finite and the other zero. For electron-induced ionizations only, this means that electrons entering from the left increase in a geometric series as they move across the space-charge region. The current thus increases exponentially, while the hole current coming from the right increases only due to the extra holes produced by the electron impact ionizations. Mathematically we set $\alpha_p = 0$, $\alpha_n = \alpha$, and write

$$\frac{\partial I_n}{\partial x} = \alpha I_n(x) = -\frac{\partial I_p}{\partial x} \tag{5.10}$$

which has the solution

$$\begin{aligned} &I_n(x) = I_n(0)e^{\alpha x}; \; I_n(w) = I_n(0)e^{\alpha w} = I \\ &I_p(x) = I - I_n(x) \end{aligned} \tag{5.11}$$

where we have set $I_p(w) = 0$, thus assuming that dark or saturation current is negligible and that photoionization occurs *at or near the left-hand boundary* and produces the current $I_n(0)$. The last equation for the hole current obeys both Eq. (5.10) but also the continuity requirement that the sum of the electron and hole currents is constant and equal to the total current I. The behavior of the two carrier currents is shown in Figure 5.11. The gain for a photoinjected electron is defined as $M = e^{\alpha w}$, and in the figure, it has a value of 10. As the junction bias voltage increases, the gain and the current increase exponentially with increasing α, which in turn increases rapidly with electric field, however there is never a "breakdown" or unlimited current increase.

The assumption of only one finite ionization coefficient is not only the simplest case, but as we shall see later, results in the best noise performance. Although we could treat any ratio, α_p/α_n, the limiting case of $\alpha_p = \alpha_n$ is not only straightforward but also applicable to many semiconductors. Here, we set $\alpha_p = \alpha_n = \alpha$ and write from Eq. 5.9,

$$\frac{\partial I_n}{\partial x} = \alpha I_n(x) + \alpha I_p(x) = \alpha I = -\frac{\partial I_p}{\partial x} \tag{5.12}$$

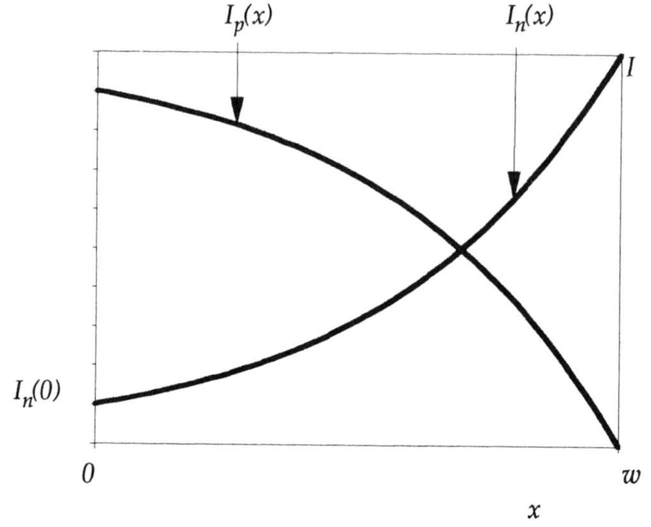

Figure 5.11 Electron and hole current variation for $\alpha_p = 0$, with photoinduced hole-electron pairs produced at $x = 0$.

with I again the total continuous current flow, which is *independent* of x. Using the same boundary conditions as previously, we obtain

$$\begin{aligned} I_n(x) &= I_n(0) + \alpha x I; \; I_n(w) = I = I_n(0) + \alpha w I \\ I_p(x) &= I - I_n(x) = I(1-\alpha x) - I_n(0) \end{aligned} \tag{5.13}$$

and the gain M becomes

$$M = \frac{I}{I_n(0)} = \frac{1}{1-\alpha w} \tag{5.14}$$

Obviously, as α approaches the quantity $1/w$, the gain becomes infinite and any small injected current will cause infinite current or avalanche breakdown. Of course, the earthly analogy would require negative mass snow or rocks moving uphill to duplicate the phenomenon. The breakdown condition, $\alpha w = 1$, simply states that each electron produces one extra hole on average, which can recross the region producing another electron, and so on. Avalanche breakdown is often the operative mechanism in Zener diodes. It occurs whenever there is a finite value for both ionization coefficients.

The behavior of the electron and hole currents is shown in Figure 5.12 with the heavy lines satisfying Eq. (5.13) with injection at $x = 0$. The light line merging with the heavy lines represents the same electron-hole pair production at an arbitrary position, $x = a$, another solution satisfying Eq. (5.12). The surprising conclusion is that the gain is *independent* of the position of the electron-hole pair creation. Therefore the good news is that photon absorption is equally effective throughout the space-charge region, rather than peaking at the p side of the junction as in the previous case. The bad news is the noise performance for the equal ionization coefficient case, discussed later, as well as the frequency response. In the single ionization coefficient case, the frequency response is the same as that of a nonavalanching diode, while with equal ionization coefficients, the response is *decreased* by M, the gain, since an injected carrier pulse makes the order of M back- and-forth transits of the avalanche region in order to attain its final gain *(Emmons, 1967)*.

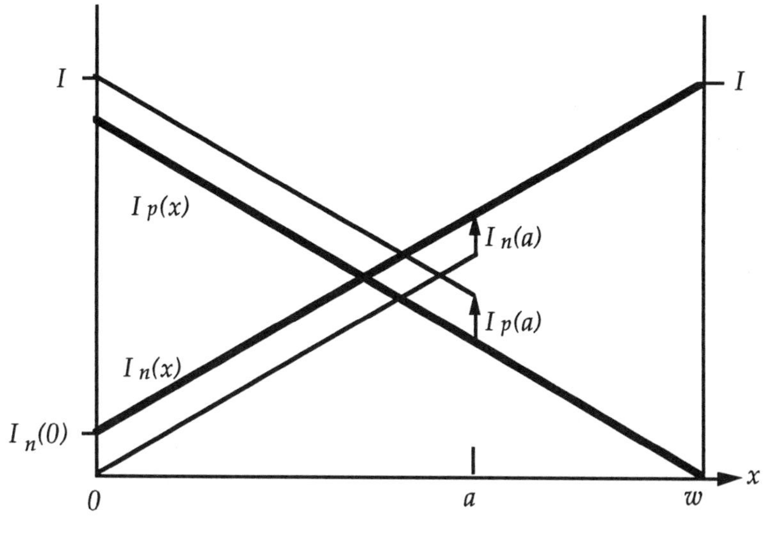

Figure 5.12 Electron and hole currents for equal ionization coefficients. The heavy line is for injection at $x = 0$. The light lines are for the same electron-hole pair injection at $x = a$.

To treat the noise, we use a procedure similar to that used for the photomultiplier. Returning to the single ionizing particle case with exponential gain $e^{\alpha w}$, we take the mean square shot noise contribution for the initial photoinduced current and add to it the shot noise of the added current and then weight these contributions by the square of the gain between the region of occurrence and the output. This of course assumes that the multiplication process is Poisson as in the photomultiplier. Mathematically, we write

$$\begin{aligned}
\left(\overline{i_n^2}\right)_{out} &= M^2[2qI_n(0)B] + \int_{I_n(0)}^{I} M^2(x)[2qdIB] \\
&= M^2[2qI_n(0)B] + \int_{I_n(0)}^{I}\left(\frac{I}{I_n(x)}\right)^2 2qBdI \\
&= M^2[2qI_n(0)B] + 2qB\left[\frac{-I^2}{I_n(x)}\right]_{I_n(0)}^{I} \\
&= M^2[2qI_n(0)B] + 2qB\left[-\frac{I^2}{I} + \frac{I^2}{I_n(0)}\right] \\
&= \left[M^2 - M + M^2\right][2qI_n(0)B] = M^2\left[2 - \frac{1}{M}\right][2qI_n(0)B]
\end{aligned} \tag{5.15}$$

where $M(x)$, the current gain between x *and* w, is given by $I(w)/I(x) = I/I(x)$. Defining F as the avalanche photodiode (APD) noise factor, we then write

$$\overline{i_n^2} = 2qFM^2I_n(0)B \qquad F = (2 - \frac{1}{M}) \tag{5.16}$$

the equivalent of Eq. (5.3) for the photomultiplier. The mean-square noise current thus behaves like that of the photomultiplier, except that the randomness in the impact ionization process produces an extra factor of two. This factor is to be compared with the Γ-factor for the photomultiplier, which is closer to unity. A simple explanation for the factor of two in the APD case is to consider the impact ionization process as the equivalent of a set of dynodes with a secondary emission ratio of $\delta = 2$. A long cascade of these "dynodes" would then yield an excess noise factor from our photomultiplier treatment of $\Gamma = \delta/(\delta - 1) = 2$. Unfortunately, with the exception of silicon, most semiconductors have ionization coefficients of roughly equal value. In addition, inherent nonuniformities in the fabrication process only allow gains up to the order of *200*. At higher gains, small regions of the junction area break down, due to the always finite ratio, α_p/α_n.

For equal ionization coefficients, we again use the treatment of Eqs. (5.15), but with $M(x) = M$, a constant, from Eq. (5.14), resulting in

$$\begin{aligned}\left(\overline{i_n^2}\right)_{out} &= M^2[2qI_n(0)B] + \int_{I_n(0)}^{I} M^2(x)[2qdIB] \\ &= M^2[2qI_n(0)B] + \int_{I_n(0)}^{I} M^2[2qdIB] \\ &= M^2[2qI_n(0)B] + M^2[2qIB - 2qI_n(0)B] \\ &= M^2[2qI_n(0)B] + M^3[2qI_n(0)B] - M^2[2qI_n(0)B] \\ &= M^3[2qI_n(0)B]\end{aligned} \tag{5.17}$$

The noise factor F in Eq. (5.16) thus becomes $F = M$, as a result of the full multiplication of the added shot noise, independent of its position in the active region.

For the general case *(McIntyre, 1966)*, if the *ratio* of the coefficients is independent of electric field, the noise factor becomes

$$F = \left(\frac{\alpha_p}{\alpha_n}\right)M + \left(2 - \frac{1}{M}\right)\left(1 - \frac{\alpha_p}{\alpha_n}\right) \tag{5.18}$$

for the case of *electron injection at the p-side of the junction*, and is plotted in Figure 5.13. A dual equation holds for hole-injection at the *n*-side. The contribution of dark current to the noise depends critically on the place of origin of the current. In the simplest case of the single ionization coefficient APD, the electron component of the saturation current originating in the *p*-type material produces the same mean square noise current as that of the photoinjected electrons at the *p*-side. In contrast the hole component originating in the *n*-type material is nonmultiplied and makes the same contribution as that in a non-avalanching diode. Thermally generated carriers in the avalanche region produce a noise be-

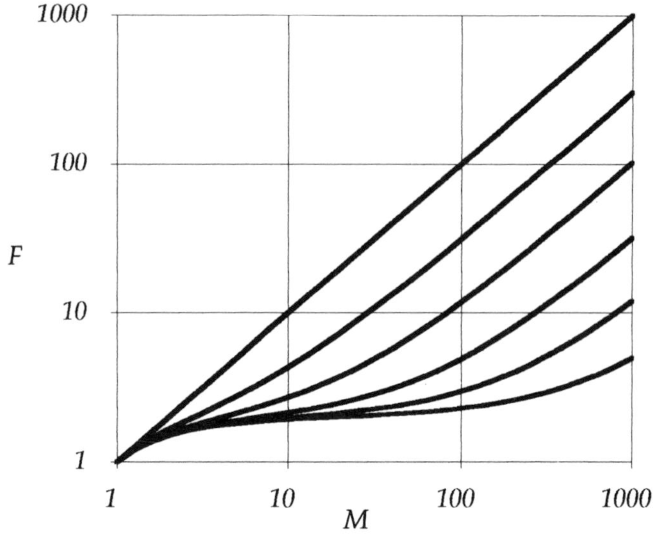

Figure 5.12 Noise factor, F, as a function of the gain, M, for electron injection at p side of the avalanche region. Ionization ratios, α_p/α_n are, from bottom, $0.003, 0.01, 0.03, 0.1, 0.3, 1$.

tween these two limits. Many APD specification sheets list the "multiplied" and "nonmultiplied" components of the dark current.

Silicon avalanche photodiodes can produce stable gain and increased sensitivity for gains up to about *200*, and available α_p/α_n ratios as low as *0.005* result in noise factors between 2 and 3. As pointed out earlier, slight nonuniformities in the junction can produce breakdown and excess noise in localized regions at gains higher than *200*. Most of the longer wavelength materials have ionization ratios close to unity and offer much smaller improvements in noise performance. Novel proposals to vary the energy gap as a function of distance through the junction offer the promise of tailoring the ionization ratios but these have thus far been unsuccessful.

For optimum performance of an APD, the gain is usually increased to the point at which the diode noise current is approximately equal to the effective amplifier noise. Neglecting dark current, the diode noise becomes "signal noise limited" and the mean square noise current and signal-limited *NEP* become

$$\overline{i_n^2} = 2qFM^2 i_S B; \quad (NEP)_{SL} = \frac{2Fh\nu B}{\eta} \tag{5.19}$$

and from Eq. (4.27), assuming negligible background radiation,

$$\left(\frac{S}{N}\right)_V = \frac{P_S}{\sqrt{\frac{2P_S Fh\nu B}{\eta} + \frac{1}{M^2}\left(\frac{h\nu}{\eta q}\right)^2 \frac{4kT_N B}{R_{in}}}} \tag{5.20}$$

where we note that the $(NEP)_{AL}$ has been reduced by the gain factor M. In any specific application the signal dependence of the noise complicates the optimization of the gain for best performance. For example, as we discuss in Chapter 7, signal-induced noise is zero when transmitting a *"0"* in a direct-detection digital communication system, but a maximum when transmitting a *"1."* It is still instructive, however, to calculate the value of NEP for a $(S/N)_V$ of unity, thus determining how close to signal-limited noise performance an APD can approach. Setting Eq. (5.20) equal to one and solving yields

$$(NEP)_{APD} = \frac{h\nu B}{\eta}\left[F + \sqrt{F^2 + \left(\frac{4kT_N}{q^2 R_{in} B}\right)\frac{1}{M^2}}\right] = F_{SL}\left(\frac{2h\nu B}{\eta}\right) \tag{5.21}$$

where we have defined a figure of merit, F_{SL}, which is the ratio of the diode NEP to the ideal signal or photon-noise-limited NEP. Using the amplifier parameters of the photomultiplier example in Section 5.1, $R_{in} = 7000\ \Omega$, $T_N = 300\ K$, and $B = 250\ MHz$, we obtain the results of Figure 5.14. The term in Eq. (5.21), $(4kT_N/q^2 R_{in} B) = 3.7 \times 10^5$ for this case. It is apparent that for all ionization ratios, there is an optimum gain for best sensitivity. The lowest ionization ratio, *0.003*, yields a performance within a factor of 3 of signal- or photon-noise-limited operation. Even

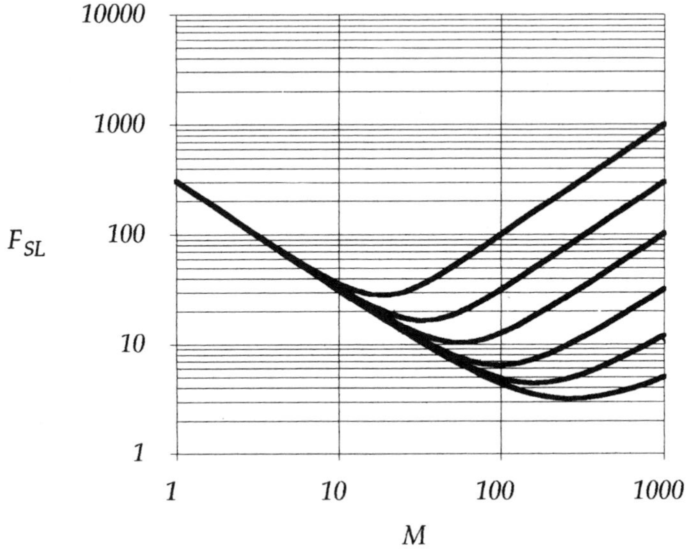

Figure 5.13. Figure of merit, F_{SL}, vs gain, M, for ionization ratios, from bottom, α_p/α_n = 0.003, 0.01, 0.03, 0.1, 0.3, 1. $R_{in} = 7000\Omega$, $T_N = 300K$, $B = 250$ MHz.

the much noisier device with unity ionization ratio shows a factor of 10 improvement from 300 down to 30 times the ideal. Note however, that the optimum gain is about 20; higher gains reduce the sensitivity. We again emphasize that this performance curve gives the optimum gain for a desired $(S/N)_V$ of unity. More general cases are treated in Chapter 7.

Example. In the actual design of any junction photodiode, there is a trade-off between quantum efficiency and transit time, the larger w is, the greater η, but the longer τ_t. Thus, guaranteeing that all electrons are photoinjected at the p-side of the junction also requires that the avalanche region be much wider than the optical absorption length, $1/\alpha_{opt}$. This in turn results in excessive transit times and poor frequency response. Let us consider the single ionization coefficient case and set the active region width equal to the optical absorption length, yielding a quantum efficiency of 63% (see Problem 5.3). We now wish to find the net gain of this structure M and its noise factor, F. Weighting the gain at position x by the fraction of photocurrent produced at that point results in

$$M = \frac{\int_0^w e^{\alpha(w-x)} e^{-\alpha_{opt} x} dx}{\int_0^w e^{-\alpha_{opt} x} dx} = e^{\alpha w}\left(\frac{\alpha_{opt}}{\alpha + \alpha_{opt}}\right)\frac{(1 - e^{-\alpha w} e^{-\alpha_{opt} w})}{(1 - e^{-\alpha_{opt} w})}$$

$$= M_{max}\left(\frac{1}{\ell n M_{max} + 1}\right)\frac{(1 - M_{max}^{-1} e^{-1})}{(1 - e^{-1})}; \quad M_{max} = e^{\alpha w} \tag{5.22}$$

where we have set $\alpha_{opt} w = 1$. As an example, if the maximum gain M_{max} is 200, then the operating gain M becomes 50. But what is the effect of this distributed photoinjection on the noise factor F? We now take the mean square noise contribution and weight it by the photocurrent fraction at the position x and then integrate, yielding

$$\overline{i_n^2} = \frac{\int_0^w M^2\left(2 - \frac{1}{M}\right) 2qIe^{-\alpha_{opt} x} B dx}{\int_0^w e^{-\alpha_{opt} x} dx}$$

$$= 2qIB \frac{\int_0^w [2e^{2\alpha(w-x)} - e^{\alpha(w-x)}] e^{-\alpha_{opt} x} dx}{\int_0^w e^{-\alpha_{opt} x} dx} \tag{5.23}$$

which, using the same definition of M_{max}, becomes

$$\overline{i_n^2} = 2qIB \times \frac{M_{max}^2}{(1 - e^{\alpha_{opt} w})}\left[\begin{array}{l} \frac{\alpha_{opt} w}{2\ell n M_{max} + \alpha_{opt} w}(1 - M_{max}^{-2} e^{-\alpha_{opt} w}) \\ - \frac{\alpha_{opt} w}{\ell n M_{max} + \alpha_{opt} w} M_{max}^{-1}(1 - M_{max}^{-1} e^{-\alpha_{opt} w}) \end{array}\right] \tag{5.24}$$

which, with $\alpha_{opt} w = 1$ and $M_{max} = 200$, becomes $2qIB$ (5400). The noise factor F is then $5400/M^2 = 5400/2500 = 2.16$, only slightly higher than the expected value for photoinjection at $x = 0$ of $(2 - 1/M) = 1.98$. These results are of course only valid for a vanishingly small hole-induced ionization coefficient. For the case of equal ionization coefficients, the gain is independent of the point of injection, and in a similar manner, the noise factor may be shown to be $F = M$ even for distributed injection across the avalanche region.

The technology and performance of APDs is still advancing. *(Wang, 1989)* gives a general description of newer APD structures, such as the SAM, with "separate *a*bsorption and *m*ultiplication" regions. He also discusses design trade-offs for minimizing dark currents especially at the longer wavelengths. In the case of silicon, the ionization coefficient values and ratio are strongly field dependent and precise control of the field profile can result in effective values of α_p/α_n as low as 0.005

Problems

5.1 An optical receiver operating at a wavelength of $1.06\ \mu m$ has an electrical bandwidth of $1000\ MHz$. Find the NEP using a high-gain PIN-FET transimpedance amplifier with an input resistance of $500\ \Omega$. The FET has a g_m of $5\ mS$, a γ of 1.5, and the quantum efficiency of the PIN is 0.5.

5.2 What would be the required quantum efficiency of a high-gain photomultiplier to yield the same sensitivity as in Problem 5.1? Assume a noise factor Γ of 1.25.

5.3 The p-i-n photodiode shown in the sketch has thin transparent p and n regions so that all photoinduced pairs are produced in the intrinsic space-charge region.

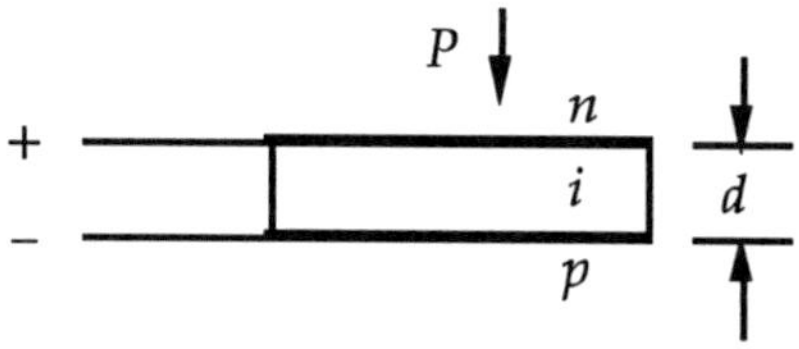

(a) Write an expression for the quantum efficiency, η, in terms of the absorption coefficient α and the thickness d.

(b) Find an expression for d *versus* α, if $\eta = 50\%$.

5.4 Find both the signal-noise-limited NEP, $(NEP)_{SL}$, and the dark-current-limited NEP, $(NEP)_{DL}$, for a photomultiplier with current gain G and excess noise factor Γ. Note that the mean square noise current is given by $2qG^2\Gamma(i_S + i_D)B$.

5.5 A photomultiplier is used to detect individual "photons" (photo-events) with a time resolution of $10\ nsec$. With a 50-Ω output load with $T_N = 300\ K$, find the required gain G for the output current pulse to have a peak value 10 times the rms noise current. For simplicity, assume a square output pulse and a bandwidth of $100\ MHz$.

5.6 A photomultiplier with a cut-off wavelength of $1\ \mu m$ is cooled to reduce the dark current. At what temperature will the $(NEP)_{DL}$ be reduced by $20\ dB\ (100/1)$ from the value at $T = 300\ K$?

5.7 A unit quantum efficiency ideal detector at a 10-μm wavelength is operated at $77\ K$ with an integral bias resistor of $1000\ \Omega$ in the same temperature bath. The detector is coupled by a cable to an external transimpedance amplifier with input resistance of $300\ \Omega$, and effective input noise current of $2\ pA/Hz^{1/2}$. Find:

(a) R_{in} and T_N as seen by the detector.

(b) The overall effective input noise current in $A/Hz^{1/2}$.

(c) The $(NEP)_{AL}$ for a bandwidth of $1\ MHz$.

5.8 A mercury-cadmium-telluride (HgCdTe) junction detector is used in the "photovoltaic" mode to detect $10\ \mu m$ radiation. The detector area is $1\ mm^2$, the measured zero bias or dark resistance is $100\ \Omega$ at $77\ K$, the operating temperature, and the bandwidth is $1\ MHz$. Find:

(a) The diode saturation current I_s.

(b) The $(NEP)_{AL}$ with zero background, with a high-gain FET stage, $g_m = 0.05\ mhos$, $\gamma = 1$, also operated at $77\ K$.

(c) The linear dynamic range, P_{max}/NEP, where P_{max} produces $i_S = I_s$.

5.9 An InGaAs APD operating at a 1.5-μm wavelength has a quantum efficiency of 0.5 and *equal impact ionization coefficients*. The diode output drives an amplifier with $R_{in} = 50\ \Omega$, $T_N = 300\ K$, and $B = 1000\ MHz$.

(a) Write an expression for the total mean square noise current in terms of the nonmultiplied signal current, i_{S0}, M, F, B, T_N, and R_{in}.

(b) Find i_{S0} for $(S/N)_V = 1$.

(c) Find and plot the NEP as a function of M from $M = 1$ to $M = 200$.

(d) Estimate NEP_{min} and the associated value of M.

Chapter 6

Heterodyne or Coherent Detection and Optical Amplification

Until this point, we have discussed optical detection in terms of the measurement of the incident *power*. With the advent of the laser, however, it has become possible to measure the *amplitude* and *phase* of the incident optical field through the process of *heterodyne detection*. This technique, used universally in modern communications and radar systems at radio and microwave frequencies, utilizes the "beating" between a single-frequency or monochromatic "local" oscillator and the received signal. In a radio receiver the "mixing" or "beating" process is carried out in a highly nonlinear rectifying element. The resultant output current at the *intermediate frequency (i.f.)* is amplified and then further processed to produce an audio, video, or digital signal output. The important property of heterodyne systems is the reproduction of the amplitude and phase of the incoming signal at a lower preselected frequency such that the signal can be band-limited for an optimum signal-to-noise ratio. Since the local oscillator frequency determines the specific frequency that will be amplified by the *i.f.* amplifier, the spectral band-pass of the heterodyne detection system is determined by the choice of local oscillator frequency and the amplifier bandwidth.

The mixing action in an optical heterodyne system does not arise from the nonlinear interaction of two electric circuit currents but is inherent in the behavior of a simple photodetector. Specifically, the detector output *current* is proportional to the input optical *power*, or the square of the *electric field*. Because of interference between two monochromatic optical waves, the instantaneous power has a sinusoidal variation at the difference or intermediate frequency. By choosing the frequency separation to occur at radio or microwave frequencies, the amplitude, phase, *and* frequency can be measured in a heterodyne optical detection system. In this sense the detection is *coherent* since the output current is a quantitative measure of the *temporal* and *spatial* relationship between the signal and local oscillator waves. A concomitant property of heterodyne detection is a receiver angular sensitivity determined by the diffraction-limited field of view of the receiver aperture. This should be contrasted with direct detection where the angular field of view is determined by the ratio of the detector dimension to the focal length and is independent of the aperture size.

In our pursuit of high-quantum-efficiency detection systems that are limited only by signal or photoevent shot noise, we have found that the photomultiplier and to a limited extent the silicon APD meet this criterion, but are unfortunately limited to wavelengths shorter than about $0.8\ \mu m$. Both these devices utilize photoelectron multiplication to overcome the noise in the following amplifier stage. An alternative to this approach is to go backward in the detection process and amplify the optical signal *before* photodetection. This may be accomplished using a laser medium such as an injection-pumped semiconductor or an optically pumped impurity-doped solid such as erbium-doped glass. The optical amplification process is also coherent since, as discussed in Chapter 2, the stimulated wave is both a directional and a phase replica of the incident wave. Because of this coherent behavior, we find it advantageous to use heterodyne concepts in our treatment of optical amplifiers. In fact, we find similar behavior to the extent that optical amplifiers operate most efficiently with a diffraction-limited field-of-view and a signal spectral bandwidth matched to the system *optical* bandwidth.

6.1 Heterodyne: Simple Plane Wave Analysis

We first treat a simple heterodyne system involving codirectional plane waves, then we extend our treatment to arbitrary wave shapes to establish the directional sensitivity of a heterodyne detection system. We consider two aligned plane waves normally incident on a flat photodetector surface, the local oscillator, with power P_{LO} and electric field E_{LO}, and the other a weak signal with power P_S and electric field E_S. The detected current is given by $i(t) = \eta q P(t)/h\nu$, where $P(t)$ is in turn given by $E^2(t)A/z_0$, where A is the photodetector area and z_0 is the wave impedance of free space. The local oscillator power at radian frequency ω_{LO} will "beat" with the signal power at ω_S yielding a sinusoidal power variation at the intermediate or difference frequency, ω_{if}. Mathematically, we may write

$$\begin{aligned} E^2(t) &= \left[E_S\cos(\omega_S t+\phi)+E_{LO}\cos\omega_{LO}t\right]^2 \\ &= \frac{1}{2}E_S^2[1+\cos 2(\omega_S t+\phi)]+\frac{1}{2}E_{LO}^2(1+\cos 2\omega_{LO}t) \\ &+ E_{LO}E_S\{\cos[(\omega_S t+\phi)+\omega_{LO}t]+\cos[(\omega_S t+\phi)-\omega_{LO}t]\} \\ &\Rightarrow \frac{1}{2}E_S^2+\frac{1}{2}E_{LO}^2+E_{LO}E_S\cos(\omega_{if}t+\phi) \end{aligned} \tag{6.1}$$

where in the last line we have dropped all terms at optical frequencies such as $2\omega_S$, etc., since the detector cannot respond at such high frequencies. Since $\omega_{if} << \omega_S$, we can use a mean frequency, $\nu \approx \omega_S/2\pi$, and write

$$\begin{aligned} i(t) &= i_S + i_{LO} + i_{if}\cos\omega_{if}t \qquad \textit{with} \\ i_s &= \eta q P_S / h\nu \\ i_{LO} &= \eta q P_{LO} / h\nu \\ i_{if} &= 2\sqrt{i_{LO} i_S} \end{aligned} \tag{6.2}$$

and we note that $i_S << i_{if} << i_{LO}$ and the *i.f.* current is proportional to the *square root* of the signal current, that is, to the optical electric field. We have here assumed that the local oscillator is *single frequency with no phase or amplitude fluctuations* so that the *i.f.* current becomes a replica of the signal field in amplitude and phase. This local oscillator criterion requires a laser source and if the laser frequency or phase fluctuates, so does that of the output signal. Ideally, however, we have performed a coherent measurement of the incoming optical signal. The noise current produced in the detection process is again shot noise but it is now produced by the local oscillator current and can be increased to such a level that it overwhelms the noise in the following resistor/amplifier system. Specifically we may write

$$\begin{aligned} &\overline{i_n^2} = 2qi_{LO}B; \quad \overline{i_{if}^2} = (2\sqrt{i_{LO}i_S})^2\overline{(\cos^2\omega_{if}t)} = 2i_{LO}i_S \\ &\left(\frac{S}{N}\right)_P = \frac{\overline{i_S{}^2}}{\overline{i_n{}^2}} = \frac{2i_{LO}i_S}{2qi_{LO}B} = \frac{i_S}{qB} = \frac{\eta P_S}{h\nu B} \end{aligned} \tag{6.3}$$

and we find that the heterodyne *NEP*, the required signal power for

$(S/N)_P = 1$, becomes $(NEP)_{HET} = h\nu B/\eta$. This is one-half the value we derived for direct or incoherent detection in the signal or photon-noise-limited case. This may be understood if we realize that the *i.f.* signal bandwidth, B, for any signal waveform is twice the bandwidth required at baseband in direct detection, because of the upper and lower sidebands about the center *i.f.* frequency in the heterodyne case. In general a pulse *envelope* of width τ at radio or optical frequencies produces a signal with a bandwidth given by $B = 1/\tau$. As we noted earlier, the theoretical heterodyne *NEP* may be achieved by sufficient local oscillator power, such that the detection system is shot noise limited. Because the *i.f.* power increases at the same rate as the noise power with increasing i_{LO}, the noise of the amplifier stage can be made negligible.

6.2 Heterodyne Detection with Arbitrary Wavefronts

Thus far, we have treated a uniform plane wave and have also neglected the relative polarizations of the signal and local oscillator waves. We now treat the general case, using the configuration of Figure 6.1, with arbitrary waves producing electric fields on the detector plane defined by

$$\begin{aligned} \mathbf{E}_S(x,y,t) &= \mathrm{Re}[\mathbf{E}_S(x,y)e^{j\omega_S t}] \\ \mathbf{E}_{LO}(x,y,t) &= \mathrm{Re}[\mathbf{E}_{LO}(x,y)e^{j\omega_{LO} t}] \end{aligned} \tag{6.4}$$

where the bold type indicates a spatial vector, and the fields $E(x,y)$ are complex. Using this formulation, we may write the signal, local oscillator, and intermediate frequency currents as

$$\begin{aligned} i_S &= \frac{\eta q P_S}{h\nu} = \frac{\eta q}{2z_0 h\nu}\int \mathbf{E}_S \cdot \mathbf{E}_S^* \, dA \\ i_{LO} &= \frac{\eta q P_{LO}}{h\nu} = \frac{\eta q}{2z_0 h\nu}\int \mathbf{E}_{LO} \cdot \mathbf{E}_{LO}^* \, dA \\ i_{if} &= \frac{\eta q}{z_0 h\nu}\int \mathbf{E}_{LO} \cdot \mathbf{E}_S^* \, dA \end{aligned} \tag{6.5}$$

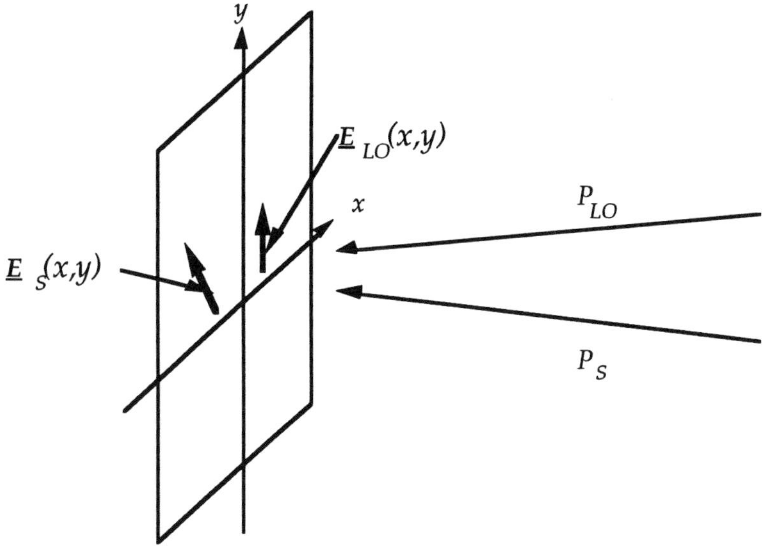

Figure 6.1 Detector plane with incident signal and local oscillator fields

which are the space-variant vector forms of Eqs. (6.1) and (6.2), with z_o the wave impedance of free space and η assumed to be independent of position in the detector plane. A more general derivation can be found in *(Kingston, 1978)*. Using Eq. (6.3), we can then write

$$\left(\frac{S}{N}\right)_P = \frac{\overline{i_{if}^2}}{\overline{i_n^2}} = \frac{\eta}{2z_0 h\nu B} \frac{\left|\int \mathbf{E}_{LO} \cdot \mathbf{E}_S^* \, dA\right|^2}{\int \mathbf{E}_{LO} \cdot \mathbf{E}_{LO}^* \, dA} \qquad (6.6)$$

and using

$$P_S = \frac{1}{2z_0} \int \mathbf{E}_S \cdot \mathbf{E}_S^* \, dA$$

we obtain the result of Eq. (6.3) for the simple plane wave case,

$$\left(\frac{S}{N}\right)_P = \frac{\eta P_S}{h\nu B}\left[\frac{\left|\int E_{LO} \cdot E_S^* dA\right|^2}{\int E_{LO} \cdot E_{LO}^* dA \int E_S \cdot E_S^* dA}\right] = \frac{\eta P_S}{h\nu B} \cdot m \tag{6.7}$$

but now multiplied by a quantity m, defined as the *mixing efficiency*, given by the bracketed term, which has a maximum value of unity if and only if the signal and local oscillator patterns are identical within a multiplicative scalar complex constant. This result may be derived using a form of the Schwartz inequality theorem. One can say that the local oscillator pattern should match the received signal pattern or in another sense, one can say that the local oscillator pattern establishes the receiver response pattern. In the simplest case, it is obvious that the signal and local oscillator polarizations must be parallel for efficient mixing because of the vector dot product in the numerator of the mixing efficiency.

6.3 Siegman's Antenna Theorem

Siegman (*Siegman, 1966*) treated the directional sensitivity of a heterodyne detector using a geometry similar to Figure 6.1. Let us assume a given local oscillator pattern, $E_{LO}(x,y)$, in the detector plane and then assume a *plane wave* signal strikes the surface, where the signal wave is given by

$$E_S(x,y) = E_o e^{jk_x x + jk_y y} \tag{6.8}$$

which represents a plane wave traveling to the *left* at an angle determined by the relative values of k_x and k_y with respect to k. If we substitute this expression into the mixing efficiency term of Eq. (6.7) we obtain a signal power response proportional to the numerator term, since the denominator is a constant for small values of k_x and k_y. Making this substitution and then taking the complex conjugate inside the absolute value brackets leads to

$$P_S \propto \left| E_O \cdot \int E_{LO} e^{-jk_x - jk_y y} dA \right|^2 = \left| E_O \cdot \int E_{LO}^{*} e^{+jk_x + jk_y y} dA \right|^2 \qquad (6.9)$$

which turns out to be the square of the Fraunhofer transform, Eq. (3.15), of the *complex conjugate* of the local oscillator distribution. But the complex conjugate of a field distribution corresponds to a time reversal and thus a wave propagating *backward* into the far field. Siegman thus showed that the receiver sensitivity pattern is the far-field pattern of a backward-propagating local oscillator beam. The elegant part of the theorem is that one can propagate the local oscillator backward through the rest of the optical system as long as *none* of the extrapolated beam is obscured by the optics. Thus, for example, the receiver power sensitivity pattern for the system of Figure 6.2 is the Airy pattern (the square of the function in Figure 3.17) produced by a uniform plane wave at the receiver system entrance. Another general theorem by *(Ross, 1970)* is called the *mixing theorem* and proves the invariance of the numerator of Eq. (6.5) over *any* surface which intercepts all signal and local oscillator power. It is also discussed in *(Kingston, 1978)*.

The result of this general analysis is the conclusion that the receiver beam shape is diffraction-limited, since it is the far-field pattern of a single plane wave propagated out from the ultimate receiver aperture.

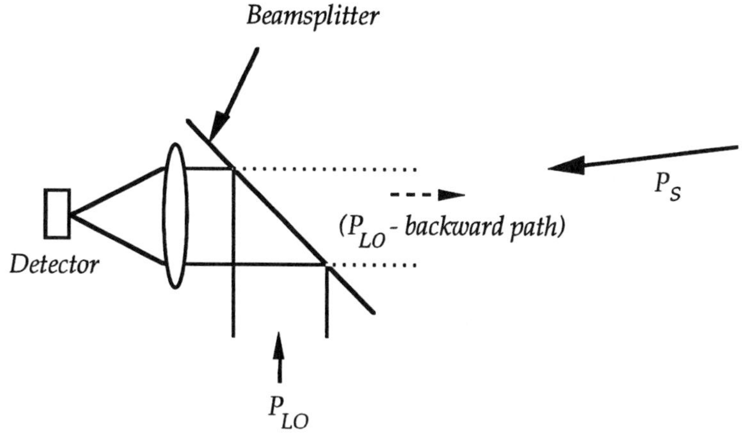

Figure 6.2 Construction to determine far-field receiver pattern local oscillator wave. The input local oscillator beam is uniform circular.

Heterodyne detection approaches photon-noise-limited performance, but the beamwidth and required frequency purity of the signal and local oscillator usually restrict its application to systems such as Doppler radar, where the frequency of the return signal must be measured.

Example. A laser radar receiver operating at a wavelength of $10\ \mu m$ uses a carbon dioxide laser as a local oscillator. The radar pulsewidth is *1 msec*, the receiver aperture diameter is *1 m*, and the required preamplifier bandwidth *1 GHz* for satellite Doppler velocity shifts, which results in a required R_{in} of $50\ \Omega$, because of the capacitance of the available junction photodiode.

Receiver beamwidth: From the antenna theorem and Section 3.4, the full beam angle at half-power for a uniform circular plane wave is $\lambda/d = 10^{-5}\ rad$ or approximately *2 arc sec*.

Local oscillator frequency stability: For optimum processing of a *1 ms* pulse, the matched bandwidth is *1 kHz*. Frequency fluctuations must be much less than *1 kHz*.

Local oscillator power: We want the local oscillator induced shot noise to be much greater than the effective amplifier input noise, $2q(\eta q P_{LO}/h\nu)B \gg 4kT_N B/R_{in}$ or $P_{LO} \gg (2/\eta)(h\nu/q)(kT_N/q)/R_{in}$. With $\eta = 0.5$ and $T_N = 300K$, $P_{LO} >> (4)(0.124)(0.026)/50 = 0.26$ *mW*.

6.4 Optical Amplification with Heterodyne Detection

In our detailed discussions of laser action in Chapters 3 and 4, we considered only the onset and development of oscillation in the laser cavity. But if the laser medium can amplify an incoming optical wave, can we use this mechanism to introduce gain in the received optical signal power prior to detection and thus improve the sensitivity? The answer is a qualified yes for a coherent or heterodyne detector but the resultant performance, dependent on the optical amplifier bandwidth, is the same as that of an unamplified heterodyne system with a detector quantum efficiency between *50%* and *100%*. For a direct detection system, where we are only measuring power, we shall find that the applicability depends on the relative magnitudes of the system bandwidth (or data rate) and the optical bandwidth of the amplifier. In addition, the noise introduced in the amplification process will be found to be proportional to the number of diffraction-limited modes amplified and then detected by a single detector. We first consider the coherent or heterodyne case.

Figure 6.3 shows a collimated signal beam passing through a laser medium, an optical spectral filter, and impinging on a heterodyne detection system to the right of the medium. The field distribution or *mode* of the beam is defined by the local oscillator beam as discussed earlier. We call this a *diffraction-limited mode,* indicating that it is the same as the radiated pattern associated with a specific field configuration on a source plane. As derived in Eq. (3.19) for a Gaussian mode, the area-solid angle product $A\Omega$ is equal to the square of the wavelength λ^2, and this relation is generally true for any type of mode (see *Kingston, 1978, p. 35 et seq.)* Such modes are *orthogonal* in the sense that the integral of the product of any two different modes over the plane is equal to zero. An example of such modes is the familiar Hermite-Gaussian set *(Boyd and Gordon, 1961),* the TEM_{mn} modes of a confocal cavity gas laser such as helium-neon. A heterodyne system only "sees" the mode, which matches that of the local oscillator in amplitude, phase, and polarization, as can be deduced from the mixing efficiency in Eq. (6.7). We use a heterodyne detection system not only to determine its overall performance but mainly as a method of calculating the noise contribution in each mode of free space.

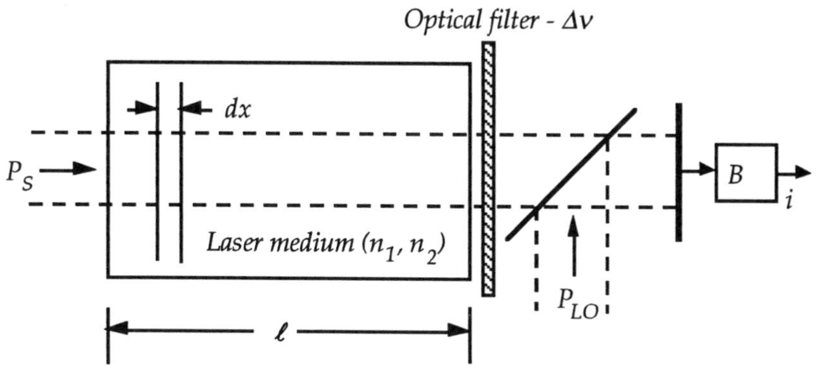

Figure 6.3 Laser amplifying medium with optical spectral filter and heterodyne detection.

As indicated in Figure 6.3, we define the laser medium in terms of the lower and upper state occupancies, n_1 and n_2, and we treat a system with simple Boltzmann statistics. First, consider the case where the medium is in thermal equilibrium and we propagate a single mode which matches that of the local oscillator. Then from the blackbody treatment of Chapter 1, the incremental power absorbed equals the incremental power emitted over a spectral bandwidth $\Delta\nu$ and is given by

$$dP = (\alpha dx)\frac{h\nu\Delta\nu}{(e^{h\nu/kT}-1)}; \quad \Delta\nu \ll \frac{kT}{h} \tag{6.10}$$

where α is the absorption coefficient and αdx therefore the emissivity of the incremental length. But the absorption coefficient is given by

$$\alpha = C(n_1 - n_2) = \frac{\lambda^2 g(\nu)}{8\pi t_S}(n_1 - n_2) \tag{6.11}$$

from Eq. (2.6). Now, since we are in thermal equilibrium with a simple Boltzmann system, the quantity $\exp(h\nu/kT)$ is the ratio (n_2/n_1) and we obtain

$$dP = C(n_1 - n_2)dx\frac{h\nu\Delta\nu}{[(n_2/n_1)-1]} = Cn_2 h\nu\Delta\nu dx \tag{6.12}$$

which states that the spontaneous emission power is proportional to n_2, the upper state population, as we would expect from Einstein's original treatment. Meanwhile let us verify the *total* spontaneous emission power into *all* modes distributed over 4π steradians of space given by the product of Eq. (6.12) and the total number of allowed modes,

$$N_{total} = 2 \times 4\pi/\Omega_{mode} = 8\pi/(\lambda^2/A) = 8\pi A/\lambda^2 \tag{6.13}$$

where the factor of two takes into account the two allowed polarizations. The resultant power becomes

$$dP_{total} = Cn_2 h\nu\Delta\nu dx(8\pi A/\lambda^2)$$
$$= \frac{\lambda^2 g(\nu)}{8\pi t_S} n_2 h\nu\Delta\nu dx(8\pi A/\lambda^2) = \frac{n_2 h\nu A dx}{t_S} \tag{6.14}$$

since, within the laser medium, $g(\nu)\Delta\nu = 1$. As expected, the total emitted power is the photon energy times the transition rate within an incremental volume Adx.

At this point we turn on the laser action and can then write the gain γ as

$$\gamma = C(n_2 - n_1) \tag{6.15}$$

and the equation for the buildup of spontaneous emission power then becomes

$$dP_n = \gamma P_n(x)dx + Cn_2 h\nu\Delta\nu dx \tag{6.16}$$

with $\Delta\nu$ now the net bandwidth including the optical filter. The first term on the right side is the growth of the amplified spontaneous emission power and the second is the new added power per unit length. Rewriting Eq. (6.16) and integrating over the length of the medium with the boundary condition that $P_n = 0$ at $x = 0$ yields

$$\int_0^{P_{n(out)}} \frac{dP_n}{\gamma P_n + Cn_2 h\nu\Delta\nu} = \int_0^{\ell} dx = \ell \tag{6.17}$$

leading finally to

$$\frac{1}{\gamma}\ln\left[\frac{\gamma P_{n(out)} + Cn_2 h\nu\Delta\nu}{Cn_2 h\nu\Delta\nu}\right] = \ell$$
$$P_{n(out)} = (e^{\gamma\ell} - 1)\frac{Cn_2 h\nu\Delta\nu}{\gamma} \tag{6.18}$$

The term $\exp(\gamma\ell)$ is the gain of amplifier G. Substituting for γ from Eq. (6.13) yields the final result

$$P_{n(out)} = (G-1)\frac{n_2}{(n_2 - n_1)} h\nu\Delta\nu \tag{6.19}$$

and we see that the noise has its lowest value for complete inversion, that is, $n_1 = 0$. As n_1 increases from zero, n_2 must increase in concert to maintain the same gain, thus increasing spontaneous emission noise. A more general treatment, taking into account the degeneracy or multiplicity of allowed upper and lower states, yields the final form,

$$P_{n(out)} = (G-1)\mu h\nu\Delta\nu \tag{6.20}$$

with the inversion factor μ given by *(Yariv, 1991)*

$$\mu = \frac{n_2}{n_2 - (g_2 / g_1) n_1} \quad \mu = \frac{n_2}{n_2 - (g_2 / g_1) n_1} \tag{6.21}$$

with g_1 and g_2 the degeneracies or densities of the respective states. Just as in the thermal or Johnson noise case in Section 4.2, the rms fluctuation in P_n is equal to the mean P_n and the output power spectral density of the amplified spontaneous emission (ASE) noise in any spatial mode becomes

$$P_{n\nu(out)} = dP_{n(out)} / d\nu = (G-1)\mu h\nu \tag{6.22}$$

Although we assumed simple Boltzmann statistics in our treatment, the results are the same for a Fermi-Dirac system such as that of a semiconductor laser, after taking into account the appropriate terms in $f(1-f)$ in the transition probabilities.

Returning to the overall detector performance, we must consider several different sources of noise. As shown in Figure 6.4, the spectral noise density of Eq. (6.22) is effective over an optical bandwidth of $\Delta\nu$, which is the *smaller* of the *optical filter* bandwidth or the *laser amplifier* bandwidth. We have assumed that the *signal* bandwidth and *i.f.* frequency are both much less than the *optical* bandwidth and now calculate the mean square *i.f.* current and the various mean square noise currents First, from Eq. (6.2),

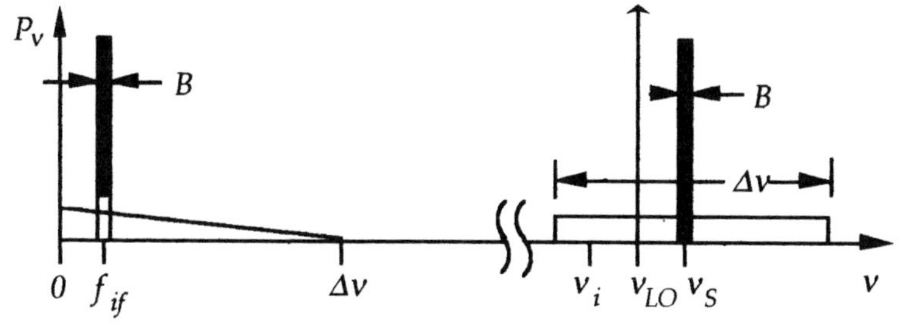

Figure 6.4 Spectral density of signal and noise powers at optical and radio frequencies. Signal and *i.f.* densities, black; noise densities, white. The single-frequency local oscillator density is a delta function.

$$\overline{i_{if}^2} = 2i_{LO}i_S = \frac{2i_{LO}\eta q G P_S}{h\nu} \tag{6.23}$$

and the dominant noise term, shown in Figure 6.4 as part of the noise in the postdetection bandwidth B, is the heterodyne frequency-shifted term,

$$\overline{i_n^2} = 2(2i_{LO}i_n) = \frac{4i_{LO}\eta q P_n}{h\nu} = 4i_{LO}\eta q(G-1)\mu B$$

Heterodyne-shifted ASE noise (6.24)

The factor of two in the second term arises from the additional "image" frequency noise produced from the local oscillator beating with noise at ν_i, a frequency f_{if} *below* the local oscillator frequency as shown in Figure 6.4.. The effective value of P_n is the spectral noise density of Eq. (6.22) times the signal or *i.f.* bandwidth B. The remaining noise terms can be made negligible if the local oscillator power is high enough, but we consider them in detail here, since some are of major significance in the direct-detection system discussed in the next section.

The mean detector currents due to the local oscillator, amplified signal, and *total* optical noise power produce shot noise giving

$$\overline{i_n^2} = 2q\left[\frac{\eta q}{h\nu}\right](P_{LO} + GP_S + P_{n(out)}) = 2q(i_{LO} + i_S + \eta q N(G-1)\mu\Delta\nu)B$$

Current-induced shot noise (6.25)

where N is the number of spatial modes seen by the detector. Finally the fluctuating component of the noise power produces a fluctuating detector current whose mean square value is

$$\left(\overline{i_n^2}\right)_{total} = \overline{\left(\frac{\eta q P_{n(out)}}{h\nu}\right)^2} = \frac{\eta^2 q^2}{(h\nu)^2} P_{n(out)}^2 = \eta^2 q^2 (G-1)^2 \mu^2 (\Delta\nu)^2 \qquad (6.26)$$

using Eq. (4.9). The spectral density of this mean square noise current falls off linearly with *electrical* frequency as shown in Figure 6.4, since the density is the self-convolution of the *optical* noise spectrum at the right of the figure. Qualitatively there are a maximum number of frequency components that can beat together to produce low frequencies and a diminishing number as the output difference frequency approaches the optical spectral width. Since the *area* of the triangle is the total mean square noise current from Eq. (6.26), the spectral density of the current noise power and the resultant circuit output for $B << \Delta\nu$ then become

$$\left(\overline{i_n^2}\right)_f = \frac{2\left(\overline{i_n^2}\right)_{out}}{\Delta\nu}\left(1 - \frac{f}{\Delta\nu}\right) \qquad \textit{Direct ASE noise} \qquad (6.27)$$

$$\left(\overline{i_n^2}\right) = \left(\overline{i_n^2}\right)_f B = 2\eta^2 q^2 \mu^2 N (G-1)^2 \Delta\nu B$$

where the quantity, N, is the number of spatial *modes* seen by the detector as a *direct power* detector. As discussed in the next section we can minimize N by an appropriate "field-stop"or focal-plane aperture. Alternatively, amplification in an active single-mode optical fiber completely eliminates noise in extra modes.

For reasonable values of i_{LO}, the heterodyne-shifted *ASE* noise becomes dominant, yielding

$$\left(\frac{S}{N}\right)_P = \frac{\overline{i_{if}^2}}{\overline{i_n^2}} = \frac{GP_S}{2(G-1)\mu h\nu B} = \frac{P_S}{2\mu h\nu B}; \qquad G >> 1 \qquad (6.28)$$

which is a factor of two lower than an unamplified heterodyne receiver with unity quantum and mixing efficiency. If the *i.f.* frequency is comparable to the optical bandwidth or if the local oscillator frequency is set near the edge of the laser gain spectrum, then the image noise can be eliminated and the ideal heterodyne performance attained. In general, unless available detector quantum efficiencies are extremely low, optical preamplification is not an attractive choice.

6.5 Optical Amplification with Direct Detection

We now consider the detection system as shown in Figure 6.5, which shows a laser preamplifier of gain G, followed by an optical filter of spectral bandwidth $\Delta\nu$, which is less than or equal to the laser amplifier bandwidth. The output radiation is focused at a field stop, which determines the field of view at the entrance to the amplifier as seen by the detector. A field stop of area A_{FS} makes the solid-angle field-of-view of the receiver equal A_{FS}/f^2, and the number of modes N seen by the detector is given by $\Omega_{FOV}A/\lambda^2$ where A is the area of the amplified beam. For $N = 1$, the diameter of the field stop would be approximately $(f/\#)\lambda$. In the case of a *single-mode* optical fiber amplifier, N is automatically unity.

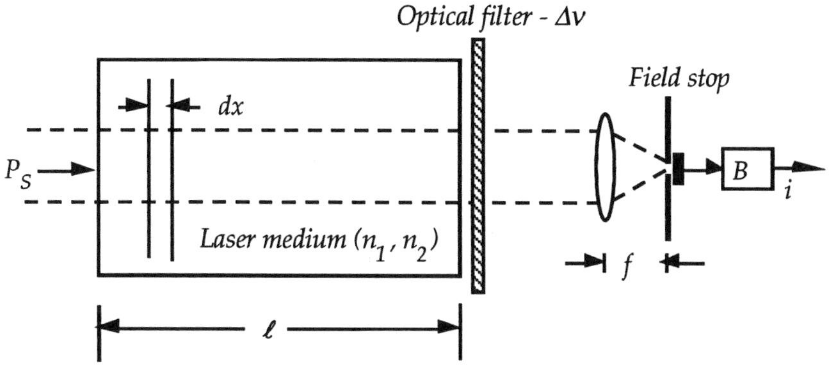

Figure 6.5 Direct detection with laser preamplification. The field stop limits the number of modes entering the detector.

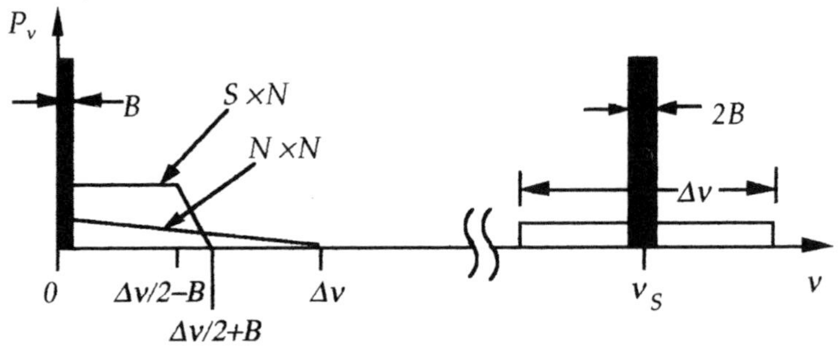

Figure 6.6 Spectral densities of signal and noise powers at optical and radio frequencies for direct detection.

As shown in Figure 6.6, we assume that the signal bandwidth is small compared with the net amplifier bandwidth, $\Delta\nu$, and note that the required signal bandwidth at optical frequencies is twice that at the detector output because of the upper and lower sidebands at optical frequencies. Since we are interested in the $(S/N)_V$ for a direct detection system, we first write the output signal current as

$$i_S = \frac{\eta q G P_S}{h\nu} \tag{6.29}$$

and then consider the several sources of noise similar to those discussed for the heterodyne case. For convenience and by practice these are separated into two classes, $N \times N$, noise "cross" noise, and $S \times N$, signal "cross" noise, as sketched in Figure 6.6. We first consider the $N \times N$ term, which is the dominant mean-square noise current produced in the *absence* of signal. From Eqs. (6.25) and (6.27), the total consists of the current-induced shot noise produced by the total mean noise power *plus* the amplified spontaneous emission (ASE) noise in the N modes striking the detector. The result is

$$\begin{aligned}\overline{i_n^2} &= 2\eta q^2 N(G-1)\mu\Delta\nu B + 2\eta^2 q^2 \mu^2 N(G-1)^2 \Delta\nu B \\ &\Rightarrow 2\eta^2 q^2 \mu^2 N(G-1)^2 \Delta\nu B; \qquad G >> 1\end{aligned} \tag{6.30}$$

with only the latter term, the $N \times N$ spectrum in Figure 6.6, important.

The $S \times N$ term is the result of mixing of the optical signal and noise fields and can be derived using heterodyne theory as in Eq. (6.24). The total mean square noise current is

$$\begin{aligned}\overline{i_n^2} &= 2 i_s i_n = 2\left(\frac{\eta q P_s}{h\nu}\right)\left(\frac{\eta q P_{n(out)}}{h\nu}\right) = \frac{2\eta^2 q^2 P_S}{(h\nu)^2}\left[(G-1)\mu h\nu\Delta\nu\right] \\ &= \frac{2\eta^2 q^2 \mu G^2 P_S \Delta\nu}{h\nu}; \qquad G >> 1\end{aligned} \tag{6.31}$$

where we have only used the noise power in the single mode which matches the signal mode because we are performing a heterodyne operation. The spectrum of this converted noise is a convolution of the optical signal and noise spectra and is sketched in Figure 6.6. For small signal bandwidths this results in a reduction of Eq. (6.31) by a factor of $B/(\Delta\nu/2)$, yielding the final result,

$$\overline{i_n^2} = \frac{4\eta^2 q^2 \mu G^2 P_S B}{h\nu} \tag{6.32}$$

Finally, there is the signal-power-induced shot noise term,

$$\overline{i_n^2} = 2 q i_S B = \frac{2\eta q^2 G P_S B}{h\nu} \tag{6.33}$$

which for high gain is negligible compared with Eq. (6.32).

The resultant $(S/N)_V$ for optical amplification with direct detection becomes, from Eqs. (6.29), (6.30), and (6.32),

$$\begin{aligned}\left(\frac{S}{N}\right)_V &= \frac{\eta q G P_S / h\nu}{\sqrt{2\eta^2 q^2 \mu^2 N G^2 \Delta\nu B + \dfrac{4\eta^2 q^2 \mu G^2 P_S B}{h\nu}}} \\ &= \frac{P_S}{\sqrt{2\mu^2 N (h\nu)^2 \Delta\nu B + 4\mu P_S h\nu B}} \\ &= \frac{P_S}{2\mu h\nu B\sqrt{\dfrac{N\Delta\nu}{2B} + \dfrac{P_S}{\mu h\nu B}}}\end{aligned} \tag{6.34}$$

For low signal-to-noise ratios, the $N \times N$ term dominates and the result reduces to

$$\left(\frac{S}{N}\right)_V = \frac{P_S}{\mu h\nu\sqrt{2NB\Delta\nu}}; \quad P_S << \mu h\nu B \tag{6.35}$$

while for large signal-to-noise ratios,

$$\left(\frac{S}{N}\right)_V = \frac{1}{2}\sqrt{\frac{P_S}{\mu h\nu B}}; \qquad \frac{P_S}{\mu h\nu B} >> \frac{N\Delta\nu}{2B} \tag{6.36}$$

which requires μ and N to be of the order of unity and $2B$, not extremely small compared with $\Delta\nu$. Under these restrictions, the amplifier produces a signal-to-noise voltage comparable to the signal-noise-limited case for an ideal detector, given from Eq. (4.26), by

$$\left(\frac{S}{N}\right)_V = \sqrt{\frac{\eta P_S}{2h\nu B}} \tag{6.37}$$

At this writing, only impurity-doped, optically pumped, fiber amplifiers satisfy the restrictions on Eq. (6.36). The use of a single-mode optical fiber reduces N to unity. A lower (n_1) state, which is normally empty and relaxes rapidly to the ground state, results in a μ also close to unity. Finally, the single-mode operation allows a narrowband optical filter, such as a Fabry-Perot cavity, whose linewidth $\Delta\nu$ is not appreciably larger than the wide bandwidth or high data-rate of $2B$ used in modern optical communication systems. We discuss the typical operating parameters of such systems in Chapter 7. Although semiconductor laser amplifiers can be operated in a single guided-wave mode, it is difficult to obtain inversion ratios adequate to produce a reasonably small value of μ.

Problems

6.1 A perfect photon detector with unit quantum efficiency is illuminated with local oscillator power, P_{LO}. Assuming unity mixing efficiency;

(a) Write a general expression for the *conversion gain*, P_{if}/P_g in terms of the load resistance, R, P_{LO}, etc.

(b) Find the numerical value of this gain for $P_{LO} = 10\ mW$, $\lambda = 1\ \mu m$, and $R = 1000\ \Omega$.

(c) Find the mean square shot noise current for $B = 100\ MHz$.

(d) For $T_N = 300\ K$, find the mean square amplifier noise current.

6.2 A photon detector is illuminated with perfectly aligned plane signal and local oscillator beams whose fields are

$$E_S(r) = E_S e^{-ar^2}; \quad E_{LO}(r) = E_{LO} e^{-br^2}$$

Find an expression for the mixing efficiency, given by

$$m = \frac{\left|\int E_{LO} E_S^* dA\right|^2}{\int E_{LO}^2 dA \int E_S^2 dA}$$

and show that it has a maximum of unity at $a = b$.

6.3 Derive Eq. (6.34) for the case of $2B = \Delta\nu$, that is, when the optical signal and filter bandwidths are equal. Note that the $S \times N$ spectrum shape becomes the same as that of the $N \times N$.

Chapter 7

Systems I: Radiometry, Communications, and Radar

To this point we have described detection systems in terms of the $(S/N)_V$ or, in the case of heterodyne detection, the $(S/N)_P$. We now apply these performance measures to several practical system problems and in the process we take into account the detailed statistics of the noise and in some cases the signal. In addition, the method of processing the signal after amplification has a significant effect on system performance. We start with radiometry, the *measurement of radiation,* and use the expected focal plane intensity to determine the output signal while the noise is determined by the detector/amplifier combination. In the case of a simple direct-detection measurement of signal intensity, the fractional fluctuation or error is the ratio of noise voltage to signal voltage or $1/(S/N)_V$. The heterodyne case is not as straightforward and here we first see the complicating effects of post-detection signal processing.

For active systems such as communications or radar, we derive relationships for bit error rates and probabilities of detection and false-alarm. Again, the postdetection processing is a critical factor as well as the choice of direct or heterodyne detection. Finally, especially with APDs or optical preamplifiers, the optical system output noise may be signal dependent and this has a marked effect on the optimal detection threshold settings and overall performance.

7.1 Radiometry

The main component of any radiometric system, whether it's thermography or television, is the optical receiver, which includes the optical element train as well as the detector. In its simplest form, as shown in Figure 7.1, the receiver can be defined by its receiver aperture area A_R, or diameter d, and focal length f. The solid angle field-of-view, Ω_{FOV}, for direct detection is given by the A_D/f^2, with A_D the detector area.

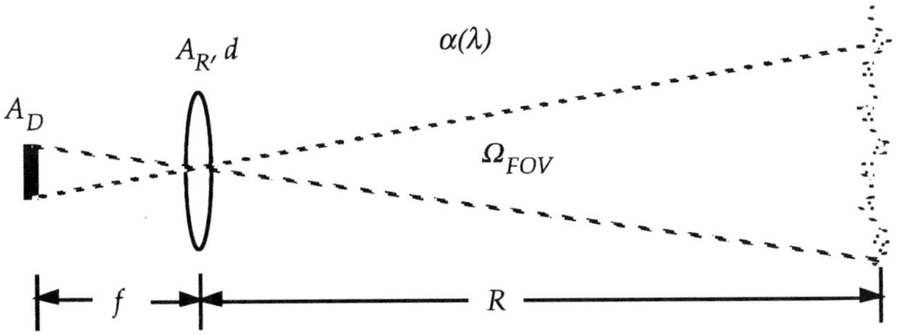

Figure 7.1 Simple optical receiver with diffuse source

For heterodyne detection, the field of view (FOV) is the solid-angle of a diffraction-limited mode, given by $\Omega_{FOV} = \lambda^2/A_R$. The range R is assumed to be much greater than the focal length and atmospheric absorption is represented by the power absorption coefficient $\alpha(\lambda)$.

For practical reasons, most optical systems contain many more elements than just a simple lens. Figure 7.2 shows three optical systems, representative of a much larger class of systems. The camera, (a), has a compound lens both for correction of "chromatic aberration," arising from refractive index variation with wavelength, and also to obtain a flat focal plane over the full angular field-of-view of the film or detector array. The "beam compressor" configuration of (b) is adequate for narrowband operation and has the advantage of producing a small beam of area A_B, which can be directed with flat reflectors to a distant detection system. Conversely, the two lenses on the right can be a movable telescope and appropriate reflecting mirrors can keep the small beam focused on the detector. The reflective system (c), also called an afocal (both foci actually at infinity) telescope, and known as a Mersenne by astronomers, has the same properties as (b), but the advantages of only one small refractive element at the detector. This attribute eliminates dispersion as well as significant lens absorption in many regions of the infrared.

We must now determine the solid-angle FOV for such systems as well as the $f/\#$ discussed in Section 1.4. This is obvious for the camera, but not so for a multielement system such as the beam compressor. For all such multielement systems we can define an *effective focal length*, f_{eff}, which is determined as follows.

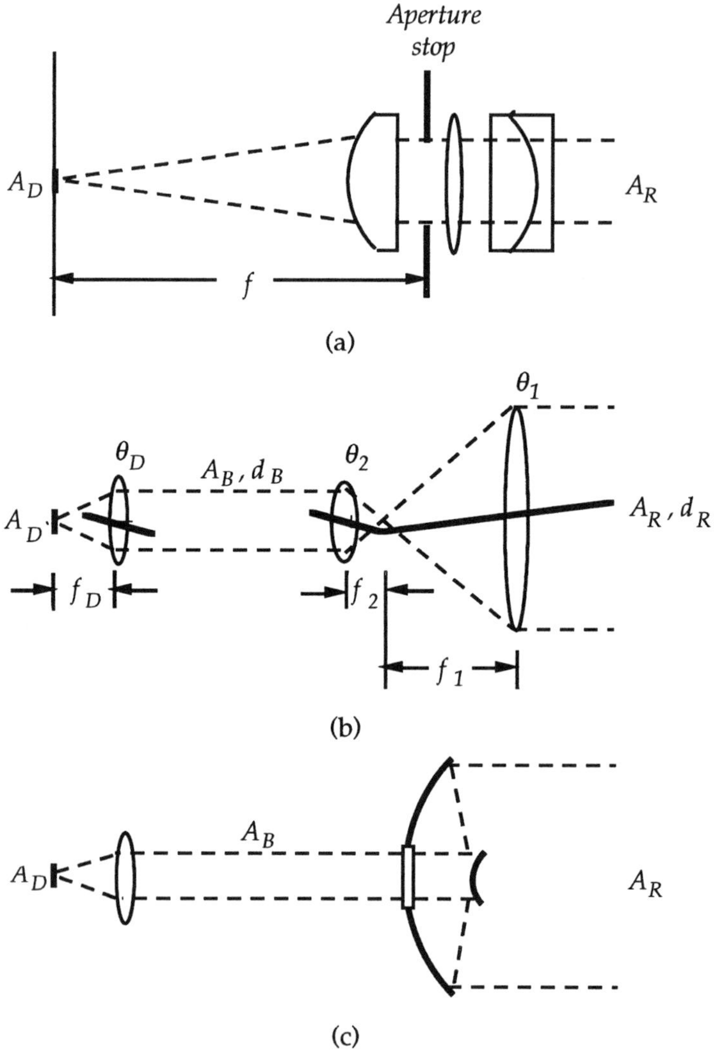

Figure 7.2 Various optical receiver configurations: (a) idealized camera, A_D is film resolution element or detector area; (b) refractive primary with beam compressor, heavy lines are ray-traces for determine effective focal length; and (c) Reflective beam compressor (Mersenne)

As in Figure 7.2(b), we determine the motion of a point on the detector plane for a small angular deviation, θ_1, of a ray entering the receiver aperture A_R. Using the rule that a ray passing through the center of a lens is undeflected, we find the displacement at the first focal point to be $f_1\theta_1$. The individual ray as drawn does not bend abruptly at this point. Rather, the conically convergent bundle of rays shifts lower and then diverges again to produce a second ray passing through the center of the second lens. The angular deviation of this ray becomes $\theta_2 = (f_1/f_2)\cdot\theta_1$ and since there is a collimated beam from the lens output to the final detector lens, the ray angle, θ_D, is the same as θ_2. Note that the angles in the figure are exaggerated for clarity. The overall *system* FOV obviously becomes limited as off-angle rays no longer strike the successive lenses in the train. The final displacement of the focal spot on the detector becomes $y = f_D\theta_2$ and we can write the effective focal length as

$$f_{eff} = \frac{y}{\theta_1} = \frac{f_D\theta_2}{\theta_1} = f_D \cdot \frac{f_1}{f_2} \tag{7.1}$$

The detector FOV then becomes

$$\Omega_{FOV} = \frac{A_D}{f_{eff}^2} = \frac{A_D}{f_D^2}\left(\frac{f_2}{f_1}\right)^2 = \frac{A_D}{f_D^2}\left(\frac{A_B}{A_R}\right) \tag{7.2}$$

where the last expression, derived from the fixed f/d ratio for the beam compressor also applies to the reflective system in Figure 7.2(c). An important conclusion from this formula is that the product $\Omega_{FOV}A_R$ is a constant for a given A_D, A_B, and f_D combination. Finally, the $f/\#$ becomes

$$(f/\#)_{eff} = \frac{f_{eff}}{d_R} = \frac{f_1}{f_2}\cdot\frac{f_D}{d_R} = \frac{f_1}{f_2}\cdot\frac{f_D}{d_B}\cdot\frac{d_B}{d_R} = \frac{f_D}{d_B}! \tag{7.3}$$

or the effective $f/\#$ is that of the detector focal length and the beam diameter at the detector lens and is independent of the beam compression. In an exact analysis, the $f/\#$ of the reflective system must be increased slightly to take into account the *obscuration* or blocking of the secondary reflector. Finally, referring to our relation for a heterodyne or diffraction-limited system, $\Omega A = \lambda^2$, we may write

$$\Omega_{FOV} A_R = \lambda^2 = \frac{A_D}{f_D^2} A_B. \quad \therefore (A_D)_{eff} = \frac{\lambda^2 f_D^2}{A_B} = \lambda^2 \frac{4}{\pi}\left(\frac{f_D}{d_B}\right)^2 \cong \lambda^2 (f/\#)^2 \tag{7.4}$$

which is consistent with the size of the diffraction-limited focal spot derived in Chapter 3.

For an extended or *resolved* source, where *the detector* FOV *is smaller than the object*, the intensity at the detector plane, I_i, is given by

$$I_i = \frac{P_i}{A_i} = \frac{P_S}{A_S} \times \frac{A_L}{\pi f^2} = H_S \frac{d^2}{4f^2} = \frac{H_S}{(2f\#)^2} \tag{7.5}$$

which is Eq. (1.24) of Chapter 1. The source in this case could be either a radiating object where the radiance H_S is determined by blackbody radiation or a diffuse reflector illuminated by the sun or some incandescent source. In either case, Eq. (7.1) assumes diffuse or "Lambertian" behavior, that is, that the reflected or radiated power varies as the cosine of the angle from the normal. For *a source smaller than the detector* FOV, which we call *unresolved*, the intensity must be reduced by the ratio of the source area to the area seen by the detector, ΩR^2. Using $\Omega = A_D/f^2$, we obtain for the signal powers,

$$\begin{aligned} P_S &= e^{-\alpha R} H_S \frac{d^2}{4f^2} A_D && \textit{resolved; direct detection} \\ &= e^{-\alpha R} H_S \frac{\lambda^2}{\pi} && \textit{resolved; heterodyne detection} \\ &= e^{-\alpha R} H_S \frac{d^2}{4R^2} A_{Source} && \textit{unresolved} \end{aligned} \tag{7.6}$$

where we have also included the effect of the attenuation of the atmosphere $\alpha(\lambda)$ and used Eq. (7.4) for the heterodyne case. It is significant that the aperture diameter does not affect the heterodyne performance in the resolved case and that the effective source area is the order of λ^2, a rather small value. This behavior results in serious performance limitations for heterodyne detection of diffuse sources.

Turbulence is also a critical parameter in many measurements. In direct-detection systems, angle-of-arrival fluctuations can cause loss of signal if these fluctuations are comparable to or larger than the detector-limited field of view. In heterodyne detection, spatial distortion of the wavefront can seriously degrade the mixing efficiency. A simple discussion of these effects is given in *(Kingston, 1978)*, and a more recent and general discussion may be found in *(Accetta and Schumaker, 1993, Vol. 2)*.

Both direct and heterodyne detection are used for radiometry in such applications as thermography, spectroscopy, and remote sensing. In some applications, such as high-resolution spectroscopy or the detection of scattered laser radiation, heterodyne detection offers superior sensitivity, especially if the source subtense is smaller than the *diffraction-limited* field of view. Determination of the available measurement precision of a heterodyne system is complicated by the detection and processing of the *i.f.* signal current and we here consider the measurement error in both direct and heterodyne detection.

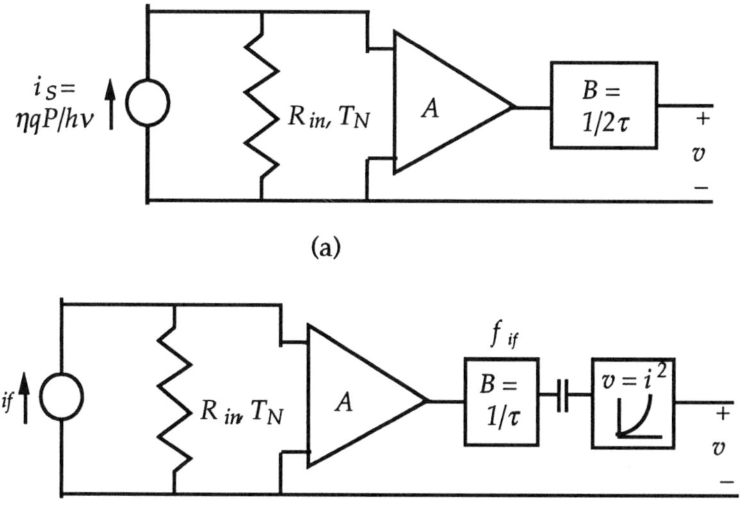

Figure 7.3 Optical signal power measurement: (a) direct detection (b) heterodyne detection.

Since we wish to measure the *power* or *energy* of the optical signal we consider the two circuits of Figure 7.3, one for direct detection, the other for heterodyne. In (a), the direct-detection circuit, the output voltage is a direct measure of the power and its integral over time, the energy. In the heterodyne circuit, (b), the *electrical* power at the *i.f.* frequency is a measure of the *optical* power and to convert this to a measurable quantity, we must pass the current through a square-law detector thus producing a voltage equal to electric power and thus proportional to the optical power. Again, the integral over time becomes the energy. In the direct detection case, the fractional rms fluctuation in the measured power becomes

$$\frac{\sqrt{\overline{(\Delta P)^2}}}{\overline{P}} = \frac{\sqrt{\overline{i_n^2}}}{i_S} = \frac{\sqrt{\overline{v_n^2}}}{\overline{v}} = \frac{1}{(S/N)_V} \tag{7.7}$$

The bandwidth B to be used in the calculation of $(S/N)_V$ and determined by the desired sampling time τ, is given by $1/2\tau$ for the direct detection case. It is instructive to calculate the fluctuation in an optical energy measurement under the ideal conditions of signal- or photon-noise-limited detection from Eq. (6.37). Here we multiply the rms power fluctuation by the sampling time τ, obtaining the rms energy fluctuation as

$$\begin{aligned}\sqrt{\overline{(\Delta U)^2}} &= \sqrt{\overline{(\Delta P)^2}}\cdot\tau = \frac{P_S\tau}{(S/N)_V} = \frac{P_S\tau}{\sqrt{\eta P_S/2h\nu B}} \\ &= h\nu\sqrt{\frac{1}{\eta}\cdot\frac{P_S\tau}{h\nu}} = h\nu\sqrt{\frac{n}{\eta}} \Rightarrow h\nu\sqrt{n};\quad \eta\Rightarrow 1\end{aligned} \tag{7.8}$$

where n is the total number of photons received during the sampling time. With unity quantum efficiency, the result is as we expected from Poisson statistics.

The heterodyne calculation is complicated by the required square-law detection. The output voltage fluctuation is not simply due to the noise term since there is an $S \times N$ term similar to that in the optical amplifier detection treatment. We start by writing the *i.f.* current including the local oscillator shot noise, which we shall assume to be much greater than the amplifier noise. In terms of the *i.f.* radian frequency ω, the total current entering the square-law detector becomes

$$i(t) = 2\sqrt{i_{LO} i_S}\cos\omega t + i_n \cos\omega t + i_n \sin\omega t; \quad \overline{i_n^2} = 2qi_{LO}B \tag{7.9}$$

where the i_n's are independent Gaussian-distributed random variables with zero mean. Their time-variation is determined by the bandwidth. This bandwidth B is, in turn, $1/\tau$, the spectral width of an *i.f.* signal sampled for a time τ. The mean output voltage from the square-law detector becomes

$$\overline{v} = \overline{i^2(t)} = \overline{\left[4i_{LO}i_S + 2i_n\sqrt{i_{LO}i_S} + i_n^2\right]\cos^2\omega t} + \overline{i_n^2 \sin^2\omega t} \tag{7.10}$$

where we have dropped the $\cos\omega t \sin\omega t$ terms, which have a timeaverage of zero. The average value of the $\sin^2$ and $\cos^2$ terms is $1/2$, and since the second term in brackets is a constant multiplied by a random variable with zero mean, it becomes zero, yielding

$$\overline{v} = \overline{i^2(t)} = 2i_{LO}i_S + \overline{i_n^2} = 2i_{LO}i_S + 2qi_{LO}B \tag{7.11}$$

But we wish to find the fluctuation in v, the low-frequency ouput voltage from the square-law detector, which from Eq. (4.3) is

$$\overline{(\Delta v)^2} = \overline{v^2} - (\overline{v})^2$$

Averaging the $\cos^2$ and $\sin^2$ terms in Eq. (7.10) to $1/2$ yields

$$\begin{aligned} v &= 2i_{LO}i_S + i_n\sqrt{i_{LO}i_S} + i_n^2 \\ \overline{v^2} &= \overline{4i_{LO}^2 i_S^2 + 12 i_n^2 i_{LO} i_S + (i_n^2)^2 + 4i_n (i_{LO} i_S)^{3/2} + 2i_n^3\sqrt{i_{LO}i_S}} \\ &= 4i_{LO}^2 i_S^2 + 12 q i_{LO}^2 i_S B + 8q^2 i_{LO}^2 B^2 \end{aligned} \tag{7.12}$$

The averaging removes the last two terms, which have zero mean. Since the square of the noise current, like the noise power, is exponentially distributed, the *mean square of the square* (the mean fourth power) is given by twice the square of the mean square, $(2qi_{LO}B)$, from Eq. (4.9).

Subtracting the square of Eq. (7.11) from (7.12) gives finally

$$\frac{\sqrt{\overline{(\Delta P_S)^2}}}{P_S} = \frac{\sqrt{\overline{(\Delta P_{if})^2}}}{P_{if}} = \frac{\sqrt{\overline{(\Delta v)^2}}}{2i_{LO}i_S} = \frac{\sqrt{4qi_{LO}^2 i_S B + 4q^2 i_{LO}^2 B^2}}{2i_{LO}i_S}$$
$$= \sqrt{\frac{qB}{i_S} + \frac{q^2B^2}{i_S^2}} = \frac{h\nu B}{\eta P_S}\sqrt{1 + \frac{\eta P_S}{h\nu B}} \tag{7.13}$$

using Eq. (6.2). The final fractional fluctuation in the power measurement is not simply the inverse of the heterodyne $(S/N)_P = \eta P_S/h\nu B$ except for $(S/N) << 1$. The significance of this result may be seen by calculating the rms fluctuation in the measured energy yielding

$$\sqrt{\overline{(\Delta U)^2}} = \sqrt{\overline{(\Delta P_S)^2}} \cdot \tau = P_S \cdot \frac{h\nu B}{\eta P_S}\sqrt{1 + \frac{\eta P_S}{h\nu B}}$$
$$= \frac{h\nu}{\eta}\sqrt{1 + \eta n} \Rightarrow h\nu\sqrt{n+1}; \quad \eta \Rightarrow 1 \tag{7.14}$$

which is similar to the direct-detection result of Eq. (7.8) except for the replacement of n by $n+1$. This turns out to be a manifestation of one form of the uncertainty principle, which states that $\Delta n \Delta \phi \approx 1$, that is, the product of the photon count uncertainty and the electromagnetic phase uncertainty are the order of unity. Although we have not used the phase of the optical wave in our measurement, heterodyne or coherent detection *measures* the phase and thus introduces extra noise or uncertainty into the measurement. We conclude then, that for laser sources or high-resolution spectroscopy, the spectral discrimination and photon-noise-limited performance of heterodyne detection give definite advantages over amplifier-noise-limited direct detection. An exception is for extremely small signals where the number of photons per sampling interval is much less than unity. Measurements of extremely weak signals are usually carried out by "switching" or "chopping" radiometry, where a known source is compared with the measured source at a switching rate, which allows ac amplification of the difference signal, synchronous detection, and then postdetection integration. Details for both direct and heterodyne detection may be found in *(Kingston,1978, p. 125 et seq.)*.

7.2 Optical Communication: Direct Detection

Direct detection is straightforward and we consider it first in our analysis of optical communication systems. Of course, we still use a laser as the source, since its diffraction-limited or single-transverse-mode behavior results in a narrow transmitted beam and thus high power density at the receiver in a free-space system. In the case of a fiber optic system, the laser single mode availability allows highly efficient coupling to the fiber and avoids the velocity dispersion associated with many modes or with the broad spectral bandwidth of an incoherent source. For the fiber, $P_S = e^{-\alpha R} L_T L_D P_T$, with α, the attenuation coefficient of the fiber, and L_T and L_D, the fractional coupling losses at the transmitter and detector. For the free-space system, using a coherent transmitter, we may write, using Figure 7.4,

$$\Omega_T = \frac{\lambda^2}{A_T}$$
$$P_S = e^{-\alpha R} P_T \frac{A_R}{\Omega_T R^2} = e^{-\alpha R} \frac{A_T A_R}{\lambda^2 R^2} P_T \tag{7.15}$$

since the fraction of the transmitted power received is the ratio of the receiver area to the transmitter beam area at range R. Because the receiver may see the transmitter against a bright background, we are also interested in calculating the background power. For a diffuse reflective or emissive source with spectral density, dH_λ in W/m^2-μm from Section

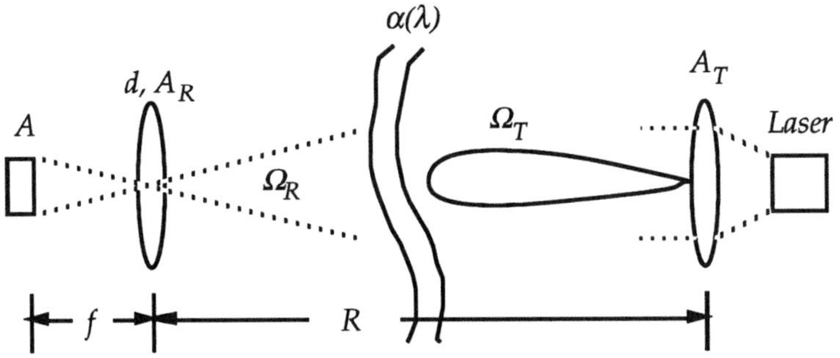

Figure 7.4 Free-space communications system.

1.5, the background power becomes

$$P_B = I_i A_D = H_\lambda \Delta\lambda \frac{d^2}{4f^2} A_D = \frac{H_\lambda \Delta\lambda A_D}{(2f/\#)^2} \tag{7.16}$$

and we note that the smaller the detector area, the smaller the background power. Conversely, the smaller the detector, the more critical the pointing precision of the receiver and the higher the susceptibility to angular fluctuations produced by the atmosphere.

We now consider two modes of direct detection; first, "photon" counting using a photomultiplier and, second, amplifier- or background-limited detection. In either case we shall use an on-off keyed signal, representing a *"1"* and a *"0,"* respectively. Each pulse is sent during a time τ, which we call the symbol time. We also calculate the BER or bit error rate, which is actually the probability of error for the digital transmission channel.

7.2.1 Photon counting

In this case the number of observed photoevents during the symbol time τ is composed of true signal events, $n_S = \eta P_S \tau/h\nu$, and noise events, $n_n = [(\eta P_B \tau/h\nu) + (I_d \tau/q)]$, where P_B is the background power and I_d is the photocathode dark current. These are of course the mean counts and the probability of a given count is determined by Poisson statistics. The probability of error or BER is one-half the sum of the probability of detecting a *1* when a *0* is transmitted, *p[1,0]*, plus the inverse error, *p[0,1]*, the probability of detecting a *0* when a *1* is transmitted. If we set a threshold count, n_T, as the minimum count for determination of a *1*, then we can write the two error probabilities as

$$p[1,0] = \sum_{k=n_T}^{\infty} p(k, n_n) \quad \text{with} \quad p(k,n) = \frac{n^k e^{-n}}{k!}$$
$$p[0,1] = \sum_{k=0}^{n_T - 1} p(k, [n_S + n_n]) \tag{7.17}$$

where the terms in the summations are the Poisson statistics of Eq. (4.2).

For a nonzero value of n_n, the results are not easy to obtain in closed form. For the special case of $n_n = 0$, the threshold n_T may be set at 1, and $p(1,0) = 0$, while $p(0,1) = e^{-n_S}$. The BER thus becomes $0.5\,e^{-n_S}$. The often-used criteria for reliable digital communications, which do not use error-correcting codes, is a BER of 10^{-9}. In this case, n_S becomes 20. For a quantum efficiency of 0.3, this would require a received pulse of 67 photons, or for the binary case considered, an average energy of 33 photons/bit. More complicated signaling schemes may use pulse-position-module, *PPM*, where there are 2^M time slots per symbol, allowing M bits per symbol. The BERs can be calculated using these principles. Unfortunately, photomultipliers are only efficient at visible wavelengths and at the longer wavelengths, the additive noise is much greater, and must be treated using large sample Gaussian statistics. Of particular significance is the existence of extremely low fiber optic attenuation only at wavelengths of $1.3\ \mu m$ and longer, where photomultipliers are effectively nonexistent.

7.2.2 Amplifier and background noise limited

In this case we consider a pulse, representing a 1 in the presence of additive Gaussian noise as shown in Figure 7.5. In (a), we show a sample time sequence of the waveform and in (b), the associated probability distribution. If the pulse height is i_S and the threshold for indicating a 1 is set at $i_T = i_S/2$, then we may write the appropriate probabilities of error as

$$p(1,0) = \frac{1}{\sqrt{2\pi \overline{i_n^2}}} \int_{i_T}^{\infty} e^{-i^2/2\overline{i_n^2}}\, di$$

$$p(0,1) = \frac{1}{\sqrt{2\pi \overline{i_n^2}}} \int_{-\infty}^{i_T} e^{-(i-i_S)^2/2\overline{i_n^2}}\, di \tag{7.18}$$

The probability of error is seen to be symmetric from the diagram and decreases with increasing signal current. The resultant error probability, or BER, is given by

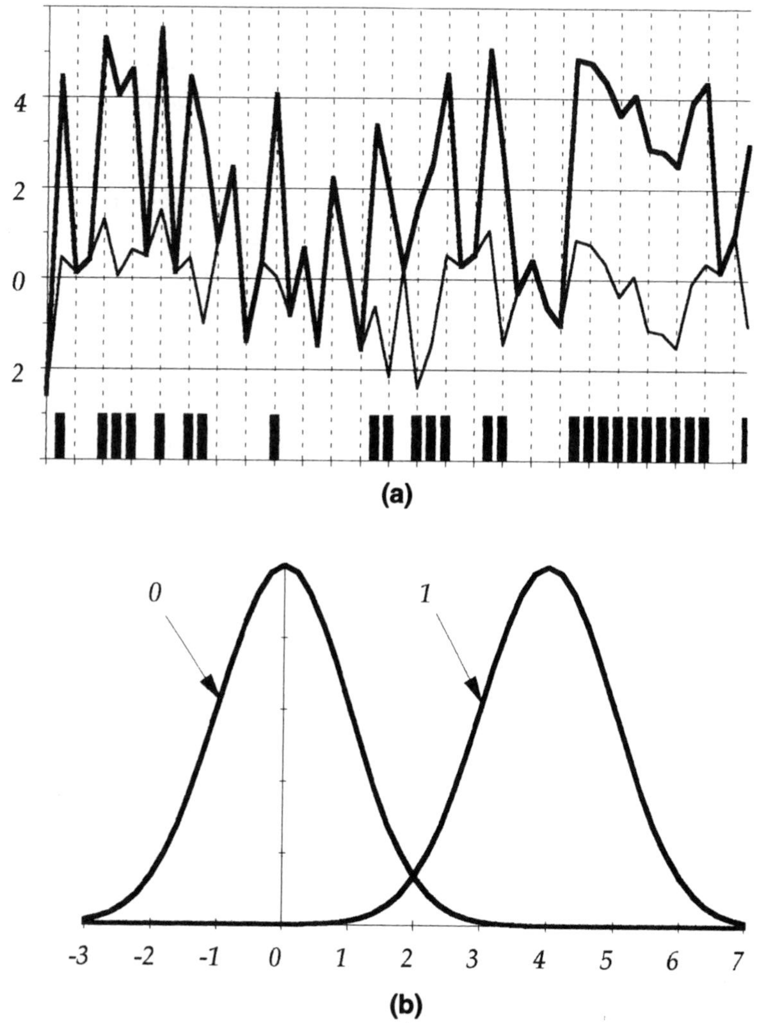

Figure 7.5 (a) Random set of signal pulses of amplitude, *4*, plus compute-generated Gaussian noise (light line) with rms value of *1*. (b) associated probability distributions; the threshold is 2.

$$BER = \frac{1}{\sqrt{2\pi \overline{i_n^2}}} \int_{i_T}^{\infty} e^{-i^2/2\overline{i_n^2}} di = \frac{1}{\sqrt{2\pi}} \int_{[(S/N)_V/2]}^{\infty} e^{-x^2/2} dx \quad (7.19)$$

and the last term is seen to be one minus the cumulative normal distribution function with argument, $(S/N)_V/2$. The result is plotted in Figure 7.6. A typical value for the *BER is* 10^{-9}, which requires a ratio of *12/1* for $i_S / \sqrt{\overline{i_n^2}} = P_S/NEP$. But P_S is the peak power, so the average power $\overline{P_S}$ is given by *6 NEP*. In the example shown in Fig. 7.5, the $(S/N)_V$ = *4*, the threshold is 2, and the predicted *BER* from Figure 7.6 is *0.02*. A close examination of Figure 7.5 indicates *4* errors, (*2* false *1's* and 2 false *0's*) or a *BER* of *4* in *50* equal to *0.08*. Although not exactly the predicted result, this is not surprising for such a small sample.

As an example of direct detection, a typical commercial PIN-FET amplifier might be specified with a *sensitivity* of *-43 dBm* for a data rate of *140 Mb/sec* at a wavelength of *1.3 to 1.5 μm*, the region of low dispersion and loss for optical fibers. The *sensitivity* is *defined* as the average power required for a BER *of* 10^{-9}. The quantity, *-43 dBm* stands for a power *43 dB* smaller than a *milliwatt* or 5×10^{-8} *W*. Using a transmitter of *1 mW* average power and state-of-the-art fiber with attenuation of *0.2 dB/km*, results in a range of *(43/0.2)* or *215 km*! The PIN-FET amplifier may typically have a noise temperature T_N of about *50 K*, a bandwidth of *90 MHz*, and an input resistance R_{in} of about *7000 Ω*,

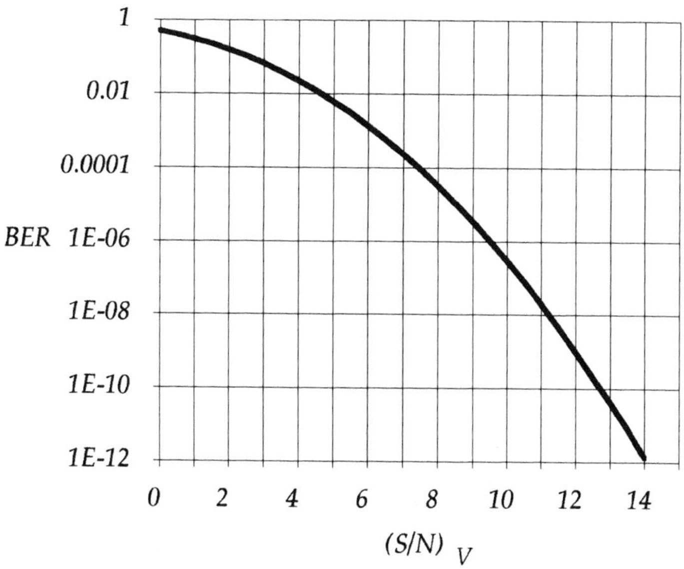

Figure 7.4 Bit error rate as a function of $(S/N)_V = P_S / NEP$

corresponding to a total photodiode-*FET* shunt capacitance of *0.3 pf*, obtained in an integrated or hybrid package. The *NEP* would then be

$$NEP_{AL} = \frac{h\nu}{\eta q}\sqrt{\frac{4kT_N B}{R_{in}}}$$

$$= \frac{1.24}{0.6(1.5)}\sqrt{\frac{4(.026)(50/300)(1.6\times10^{-19})(90\times10^{6})}{7000}} = 0.8\times10^{-8}\ W$$

at a wavelength of $1.5\ \mu m$, and an assumed quantum efficiency of *0.6*. These assumed amplifier parameters give excellent agreement with the expected required sensitivity of *6 NEP*. The received optical power of $5\ \times 10^{-8}\ W$ at a data rate of *140 Mb/s* corresponds to approximately *2500 photons/bit*, markedly higher than the photon-counting case already discussed.

7.2.3 Direct detection with signal-dependent noise

We have thus far treated the two extremes of signal-limited noise and amplifier-limited noise for the photomultiplier and PIN-FET, respectively. In the latter case, at wavelengths where photomultipliers are inoperative, signal-independent additive noise dominates the detection process. $(S/N)_V = P_S/(NEP)_{AL}$, and the BER may be calculated from Figure 7.6. With the advent of low noise-factor avalanche photodiodes (currently available only at wavelengths of less than $1\ \mu m$) and optical amplifiers, we must treat the more complicated problem of combined additive and *multiplicative* or signal-induced noise. In the *APD* case, the two noise terms are given in the denominator of Eq. (5.20), and for the optical amplifier, Eq. (6.34).

We start by defining the signal to noise, *in the presence of signal*, as $(S/N)_V$ and the ratio of rms noise currents *without* and *with* signal as

$$\delta = \sqrt{\overline{i_{n0}^2}}\,/\sqrt{\overline{i_{n1}^2}} \tag{7.8}$$

where the subscript *0* applies to the transmission of a "zero," and *1* to a "one". An example of the associated probability distributions is given in Figure 7.7, for $(S/N)_V = 3$ and $\delta = 0.5$. For equal probability of zeroes and ones, we must set the threshold so that the error probabilities are

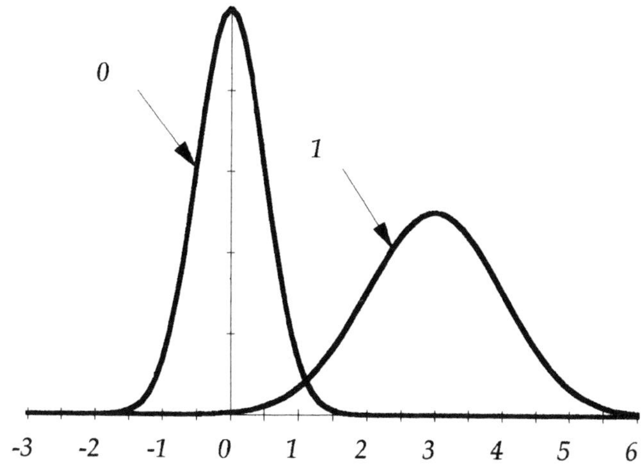

Figure 7.7 Probability distributions with signal-dependent noise. The rms noise is *1.0* with signal and *0.5* without signal. In this case, $(S/N)_V = 3$ and $\delta = 0.5$.

equal as given by

$$p(1,0) = \frac{1}{\sqrt{2\pi \overline{i_{n0}^2}}} \int_{i_T}^{\infty} e^{-i^2/2\overline{i_{n0}^2}} di \quad \Rightarrow \frac{\sqrt{\overline{i_{n0}^2}}}{i_T \sqrt{2\pi}} e^{-i_T^2/2\overline{i_{n0}^2}}; \quad i_T >> \sqrt{\overline{i_{n0}^2}}$$

$$p(0,1) = \frac{1}{\sqrt{2\pi \overline{i_{n1}^2}}} \int_{-\infty}^{i_T} e^{-(i-i_S)^2/2\overline{i_{n1}^2}} di \Rightarrow \frac{\sqrt{\overline{i_{n1}^2}}}{(i_S - i_T)\sqrt{2\pi}} e^{-(i_S-i_T)^2/\overline{i_{n1}^2}};$$

$$(i_S - i_T) >> \sqrt{\overline{i_{n1}^2}} \tag{7.21}$$

For error probabilities of less than *0.001*, corresponding to current ratios at the right greater than *3*, the approximation shown is within *10%* or less of the exact value. In this limit then, setting the two probabilities equal yields

$$\frac{i_T}{\sqrt{\overline{i_{n0}^2}}} = \frac{(i_S - i_T)}{\sqrt{\overline{i_{n1}^2}}} \quad \therefore \frac{i_T}{\delta} = (i_S - i_T); \; i_T = \frac{\delta}{(\delta + 1)} i_S \tag{7.22}$$

and the resultant *BER* is shown in Figure 7.6 for several values of δ.

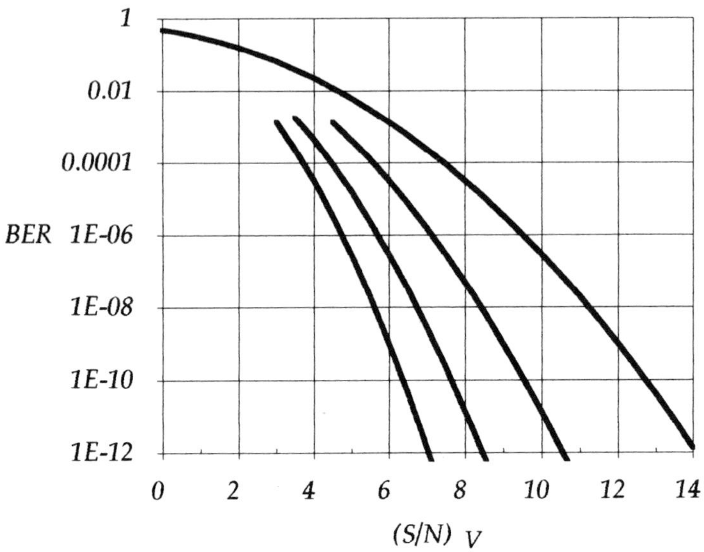

Figure 7.8 Bit error rates with signal-dependent noise for several ratios of *0* noise to *1* noise, δ. From left to right, $\delta = 0, 0.2, 0.5,$ *and* 1.

Alternatively, for a BER of less than *0.001*, the required $(S/N)_V$ from Figure 7.6 can be reduced by the factor $(1+\delta)/2$, using Eq. (7.22).

Example. A single-mode *1.5 -μm* wavelength fiber optic amplifier is used in a *10 Gb/sec* binary communication system with a required BER of 10^{-9}. A Fabry-Perot optical filter with a bandwidth of *50 GHz* follows the amplifier, and the postdetection bandwidth is *5 GHz* as determined by the data rate. The appropriate expressions are obtained by combining Eqs. (6.34) and (6.36), yielding

$$\left(\frac{S}{N}\right)_V = \frac{P_S}{2\mu h\nu B\sqrt{\dfrac{N\Delta\nu}{2B}+\dfrac{P_S}{\mu h\nu B}}} = \frac{1}{2}\sqrt{\frac{P_S}{\mu h\nu B}}; \quad \frac{P_S}{\mu h\nu B} >> \frac{N\Delta\nu}{2B} \tag{7.23}$$

From Figure 7.8, we know that $(S/N)_V$ must be at least greater than *6*, so that the quantity $(P_S/\mu h\nu B) > 144$. By comparison, the quantity, $N\Delta\nu/2B = 5$, since $N = 1$ for a single-mode fiber. Thus the inequality condition in Eq. (7.11) is satisfied and δ becomes

$$\delta = \frac{\sqrt{\frac{N\Delta\nu}{2B}}}{\sqrt{\frac{N\Delta\nu}{2B} + \frac{P_S}{\mu h\nu B}}} = \frac{\sqrt{5}}{\sqrt{5+144}} = 0.18 \tag{7.24}$$

From Figure 7.8, the required $(S/N)_V$ for BER = 10^{-9} is then 7, quite close to our original estimate, thus validating the inequality used. Erbium-doped glass amplifiers have demonstrated inversion factors μ of 1.5 or less, resulting in a final required value of $P_S/h\nu B = 4\times7^2\times1.5 \approx 300$. The *average* power is then $150h\nu B$, and the energy per bit becomes $150h\nu B/2B = 75\ photons/bit$, close to the performance of the photon-counting system. Although the required $(S/N)_V$ is only reduced by a little less than 2, when taking into account the signal-induced noise, the required *signal power* is reduced by almost 4, a significant system improvement.

In closing we note that if δ actually goes completely to zero, then the threshold also becomes zero and it is necessary to revert to the earlier photon-counting case for a correct (and much more complicated) analysis. The same is true for values of BER greater than 10^{-3} where simple Gaussian statistics no longer apply.

7.3 Optical Communication: Heterodyne Detection

Since heterodyne detection allows measurement of the frequency or phase of the optical signal, it allows much more flexibility in the signal format, although the laser frequency and phase stability must satisfy fairly rigorous requirements. As a rule of thumb, any heterodyne system requires a laser linewidth or frequency fluctuation much less than the data rate, which is of course of the order of the postdetection bandwidth. Although amplitude or on-off keying is feasible, if the laser is turned on and off, serious frequency shifts or "chirps" may occur. Alternatively, external amplitude modulators are generally inefficient at high data rates and, in addition, one-half of the laser power is wasted. Following microwave and rf communications techniques, FSK or PSK (frequency or phase shift keying) are the more attractive alternatives, because the laser may operate at a steady power level and achieve much narrower linewidths. As an example, an AlGaAs-GaAs heterostructure laser can be frequency tuned at about *300 MHz/mA* of injection current, while still maintaining the order of a *10-MHz* linewidth at each frequency setting. Using such a laser for BFSK, binary frequency shift keying, we would pass the intermediate frequency output from the heterodyne detector through two separate frequency filters corresponding to a *1* or a *0*. After envelope detection or rectification, we then would pick the "*greatest of*" current, which then determines the digit. To calculate the bit error rate, we must examine the probability distribution of the output of the envelope detector, which is given by

$$p(i) = \frac{i}{\sqrt{\overline{i_n^2}}} e^{-(i^2 + i_{if}{}^2)/2\overline{i_n^2}} I_o\left(\frac{i\, i_{if}}{\overline{i_n^2}}\right) \tag{7.25}$$

a much more complicated form than the simple Gaussians used in Eq. (7.18). The term $I_o(z)$ is the modified Bessel function of zero order and Eq. (7.25) describes *Rician* statistics after *(Rice, 1954)*. Using these statistics, which are discussed in greater detail in section 7.5, one can calculate the BER as a function of the *i.f. power* signal-to-noise ratio which is equal to the optical input P_S/NEP for the heterodyne case. The resultant curve decreases with increasing signal power and then levels off to a plateau determined by the finite linewidth of the laser. For the zero-linewidth case, a BER of 10^{-2} requires $P_S/NEP = 10$; a BER of 10^{-9}, 50. Since $NEP = h\nu B/\eta$ for a heterodyne detector, and $B = 1/\tau$, the symbol period, then the number of photons/bit becomes $50/\eta$, which is comparable to the photon counter and optical amplifier performances.

All this improved performance comes at the cost of assuring narrow laser linewidth and maintaining high mixing efficiency at the detector. Heterodyne detection also requires maintenance of polarization, a difficult task in fiber optic systems, although polarization diversity or tracking can be used. In free-space systems, circular polarization is feasible but the extremely narrow receiver and transmitter beamwidths result in a requirement of extreme tracking precision.

7.4 Radar

7.4.1 Radar equation for resolved and unresolved targets

In the radar case we consider collocated transmitting and receiving apertures of effective area, A_T and A_R with a target at range R as shown in Figure 7.9 As in the passive detection case, we must consider a resolved target, that is, where the target completely fills the transmitting beam, and an unresolved target, where the target is small compared to the beam size. In the resolved case, the power reflected from a diffuse target with reflectivity ρ is $\rho P_T e^{-\alpha R}$ and, with the receiver beamwidth equal to or greater than the transmit beamwidth, the received power becomes, using Eq. (1.23),

$$P_S = P_T e^{-\alpha R} \frac{A_R}{\pi R^2} e^{-\alpha R} = \frac{\rho A_R}{\pi R^2} e^{-2\alpha R} P_T \qquad (7.26)$$

For the unresolved case, we represent the target by a *radar cross section* σ, which is that area which if scattering *isotropically* (that is, into 4π *steradians*) would produce the observed power at the receiver aperture. As an example, a *unit reflectivity specular sphere* of radius, r, has a radar cross section of πr^2 since it scatters isotropically and intercepts incident power $I_T \pi r^2$, where I_T is the transmitted intensity at the target position.

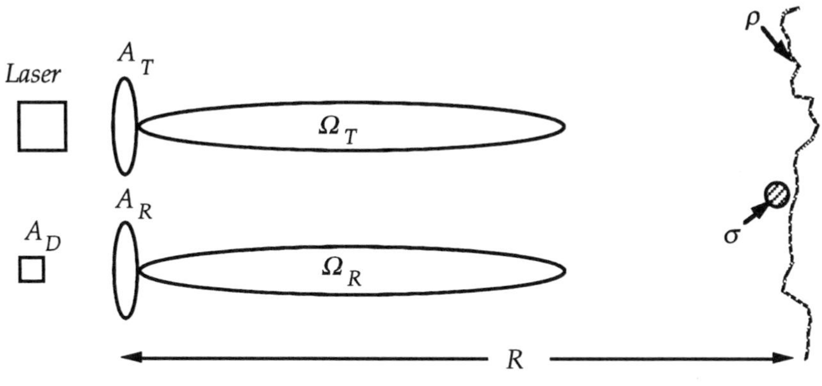

Figure 7.9 Radar system with resolved and unresolved targets.

On the other hand, a *diffuse flat plate* of area A has a cross-section of $4\rho A$ at normal incidence, since the received power is 4 times that of an isotropic scatterer from Eq. (1.23). In a similar vein, a *diffuse sphere* may be shown to have a σ of $(8/3)\pi r^2$ because of the stronger scattering back toward the transmitter. Finally, a *unit reflectivity flat plate at normal incidence* has a huge "specular" cross section of $4\pi A^2/\lambda^2$, since the reflected radiation spreads into a narrow beam of solid angle λ^2/A.

The received power in terms of this cross section becomes

$$P_S = \left(P_T \frac{\sigma}{\Omega_T R^2} e^{-\alpha R} \right)\left(\frac{A_R}{4\pi R^2} e^{-\alpha R} \right) = \frac{\sigma A_T A_R}{4\pi \lambda^2 R^4} e^{-2\alpha R} P_T \tag{7.27}$$

where we have used $\Omega_T = \lambda^2/A_T$ for the coherent transmitter beam. This is one of many forms of the radar equation and usually the transmitting and receiving areas are the same since they use the same aperture. The critical term for the unresolved target is the R^4 behavior resulting from the round-trip divergence of the transmitted and backscattered beams.

7.4.2 Performance characterization, detection and false-alarm probabilities

Just as in the communications case, noise in the receiver can produce a false decision by the signal processor. In communications a *1* may be mistaken for *0*. In a radar system, a noise peak may be mistaken for a target, yielding a *false alarm,* and we call this probability, p_{fa}. Alternatively, a negative noise peak may suppress the signal pulse below the threshold, resulting in a missed target return. By tradition, this possibility is characterized by the quantity p_d, the probability of detection, which is one minus the probability of the missed target. Unlike binary communication where the error probabilities are set equal, we want a vanishingly small error or false alarm probability along with a target detection probability usually greater than *90%*. For direct detection, the calculation can again be represented by Figure 7.5, but the threshold current, i_T is set to give a fixed value of p_{fa}, and then p_d determined from

$$
\begin{aligned}
p_{fa} &= \frac{1}{\sqrt{2\pi \overline{i_n^2}}}\int_{i_T}^{\infty} e^{-i^2/2\overline{i_n^2}}\,di = 1-\frac{1}{\sqrt{2\pi}}\int_{-\infty}^{(T/N)} e^{-x^2/2}dx;\quad \frac{T}{N}=\frac{i_T}{\sqrt{\overline{i_n^2}}} \\
p_d &= \frac{1}{\sqrt{2\pi \overline{i_n^2}}}\int_{i_T}^{\infty} e^{-(i-i_S)^2/2\overline{i_n^2}}\,di = \frac{1}{\sqrt{2\pi}}\int_{-\infty}^{([S/N]-[T/N])} e^{-x^2/2}dx
\end{aligned}
\tag{7.28}
$$

The results, shown in Figure 7.10, apply to a single pulse and thus a single measurement. For more than one pulse, averaging effects will improve the performance. For most radar systems, a p_d of *90%* with a p_{fa} of 10^{-6} yields satisfactory performance, especially with the observation of several pulse returns. From the graph, this requires a value of P_S/NEP of 6.

Heterodyne systems not only measure range and angle, but also velocity as determined by the *Doppler* shift and the resultant frequency change of the return signal. For detection and range measurement, envelope or peak detection of the *i.f.* signal is the optimum process and the detected pulses obey Rician statistics as indicated earlier for communications. We can write the *i.f.* current as

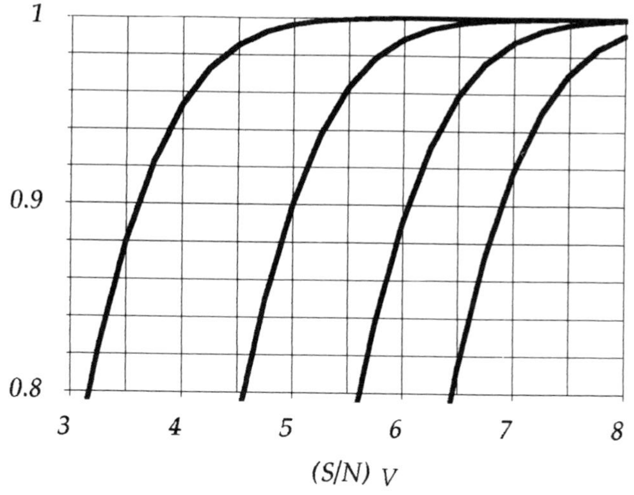

Figure 7.10 Target detection probability p_d for direct detection radar. False alarm probabilities p_{fa} from left to right: $10^{-2}, 10^{-4}, 10^{-6}$ and 10^{-8}.

$$i(t) = i_{if}\cos\omega t + i_n\cos\omega t + i_n\sin\omega t = i_I\cos\omega t + i_Q\sin\omega t \tag{7.29}$$

as we did in Eq. (7.9). Here the subscripts I and Q refer to the *in-phase* and *quadrature* or 90° phase shifted-current components. Envelope or peak detection measures the peak current averaged over many cycles of the *i.f.* frequency and is superior to square-law detection for high signal-to-noise ratios *(Davenport and Root, 1958)*. The two-dimensional plot of Figure 7.11(a) shows contours of the current distribution without and with signal. The envelope or peak current statistics are obtained by integrating the net density at a given radius i from the origin. The result, the same as Eq. (7.25), is

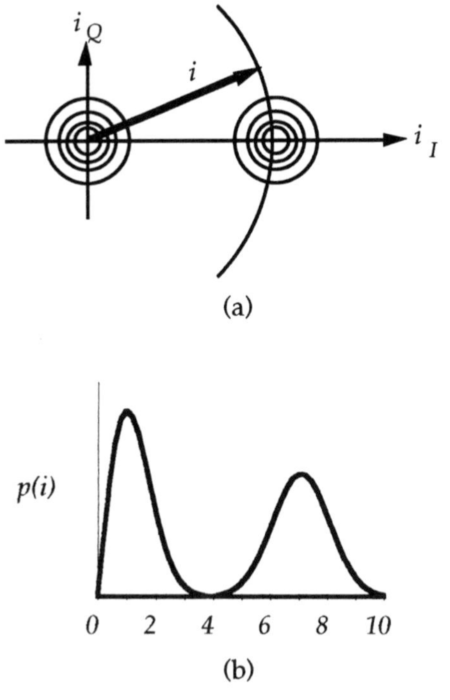

Figure 7.11 (a) Graphical representation of Rician statistics calculation. (b) Probability distribution for $i_{if} / i_n = 7$ with $i_n = 1$.

$$p(i)=\frac{i}{\sqrt{\overline{i_n^2}}}e^{-(i^2+i_{if}^2)/2\overline{i_n^2}}I_o\left(\frac{i\,i_{if}}{\overline{i_n^2}}\right)$$

and is shown in the lower plot for an *i.f* current-to-noise current ratio of *7*. In a qualitative way, the advantage of envelope detection is the second-order contribution of the quadrature noise to the *p(i)* distribution. Note that if i_Q were zero in the presence of a strong signal, there would be little change in the distribution. The main effect of quadrature noise is a phase fluctuation in the overall current, which is ignored by the envelope detector. Since heterodyne systems have an output electrical power proportional to the input optical power, the quantity $(S/N)_P$ is generally used for system description and it is usually measured in *decibels (dB)*. For example, the *i.f.*–current to noise-current ratio of *7/1* becomes a power ratio of $7^2/2 = 24.5 = (10\ \log_{10}24.5)\ dB = 13.9\ dB$. Using the form of Eqs. (7.28) with the new probability distributions, we obtain the results shown in Figure 7.12 and we note that a $(S/N)_P$ of *13.9 dB* corresponds to a *96%* probability of detection at a 10^{-6} false-alarm probability. Thus, if the radar pulse rate is *100 Hz*, we would expect a false alarm once every 10^4 *sec* or about every three minutes. The probability of a missed detection is *0.04*, and for a second pulse,

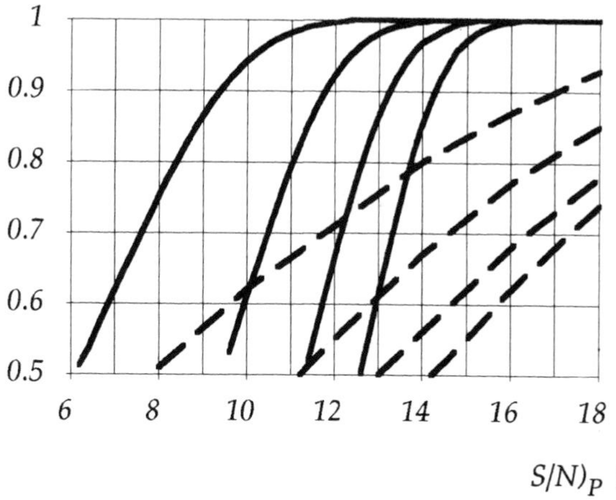

Figure 7.12 Detection probability p_d for decreasing values of $p_{fa} = 10^{-2}, 10^{-4}, 10^{-6}$, and 10^{-8}. Heavy line, constant cross-section; dashed line, Rayleigh fading cross section.

0.04. Thus, for two pulses the probability of at least one detection is $(1 - 0.04^2) = 99.8\%$.

Finally we consider the reflection from a diffuse or multiple-scattering-center target illuminated by monochromatic radiation. The scattering in this case, based on early theories of Rayleigh, has a statistical distribution because of the random addition of the electric fields from the various scattering centers. (It is this random superposition that gives the "speckle" of a reflected HeNe laser beam.) The coherence length of laser radiation is $c/\Delta\nu$, the ratio of the velocity of light to the laser linewidth. If the target depth or range extent is small compared to the coherence length, then full "Rayleigh" fluctuations are observed and the probability distribution of the observed σ becomes

$$p(\sigma') = \frac{1}{\sigma} e^{-\sigma'/\sigma} \tag{7.30}$$

where σ is the average or expectation value of the cross section and σ' the instantaneous value. The statistical distribution of the cross sections can be recognized as the power distribution for I and Q electric field components which are each Gaussian distributed. The resultant fluctuation or "fading" has a profound effect on the detection probability as indicated in the dashed lines of Figure 7.12. Again pulse averaging can markedly improve the performance, if successive pulses obtain independent samples. This will be the case if the target rotates with respect to the line-of-sight by an angle of the order of the wavelength divided by the target diameter. In this case there is a phase change across the array of random scatterers of the order of π and an independent sample is obtained.

In direct-detection radar systems, Rayleigh fading is not usually a problem since the spectrum of the laser pulse is usually broad or there are multiple lines and averaging occurs in a single pulse for a finite depth target. For example, a Nd:YAG laser at a wavelength of $1.06\ \mu m$ may have a spectral spread of the order of $30\ GHz$ resulting in a coherence length of $1\ cm$, so that target depths of many centimeters will result in relatively constant return strength.

7.4.3 Radar measurements

In addition to detection and false-alarm probabilities in a radar system, we are also interested in the measurement of such parameters as target range, angle, and velocity. By common practice we distinguish between *resolution* and *precision*. The first parameter, resolution, measures the ability of the radar to resolve or distinguish between separate targets that are close together in the measurement dimension. The second parameter, precision, yields the fluctuation or uncertainty in the measured value of a single target's position in the range, angle, or velocity dimension.

For the range dimension, $R = ct/2$, where t is the round-trip time for the pulse. We start with a direct-detection system and assume a Gaussian *received* current pulse , as shown in Figure 4.10, which has time dependence and an associated power system function given by

$$i_S(t) = i_{S0}e^{-2\pi t^2/\tau^2}; \quad |H(f)|^2 = e^{-\pi f^2\tau^2} \tag{7.31}$$

We shall somewhat arbitrarily choose the range resolution to correspond to a time separation at which two equal power pulses overlap at their e^{-1} powers, yielding

$$\frac{2\pi t^2}{\tau^2} = 1; \quad \Delta t = 2t = \frac{2\tau}{\sqrt{2\pi}}; \qquad (\Delta R)_{res} = \frac{c\Delta t}{2} = \frac{c\tau}{\sqrt{2\pi}} \tag{7.32}$$

A simple algorithm for measuring the range is to measure the position of the pulse peak by taking the time derivative of the signal current and locating the zero-crossing. The signal current out of such a differentiator then becomes

$$\frac{\partial i_S}{\partial t} = -i_{S0}\frac{4\pi t e^{-2\pi t^2/\tau^2}}{\tau^2} \tag{7.33}$$

as shown by the solid line in Fig. 7.13. But the noise out of the differentiator will produce an *rms* fluctuation represented by the dashed lines with a consequent fluctuation in the zero-crossing point. The mean square noise component is

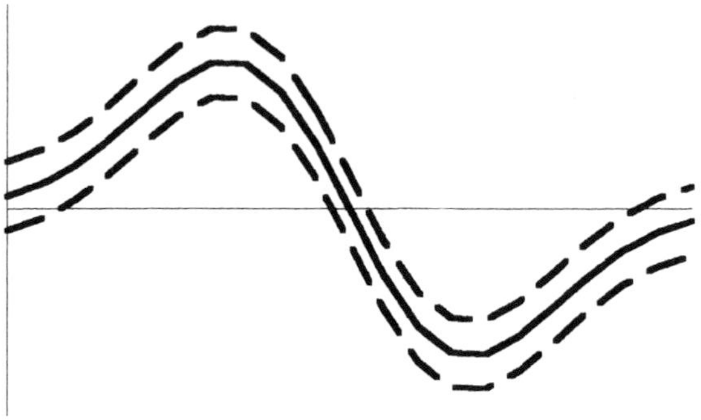

Figure 7.13 Differentiated current pulse, solid line, and rms bounds of added differentiated noise current.

$$\overline{\left(\frac{\partial i_n}{\partial t}\right)^2} = \omega^2 \overline{i_n^2} = \int_0^\infty 4\pi^2 f^2 (\overline{i_n^2})_f \left|H(f)\right|^2 df = (\overline{i_n^2})_f \int_0^\infty 4\pi^2 f^2 e^{-\pi f^2 \tau^2} df$$
$$= (\overline{i_n^2})_f \frac{\pi}{\tau^3} = \overline{i_n^2} \frac{2\pi}{\tau^2} \qquad from \;\; (\overline{i_n^2})_f = \frac{\overline{i_n^2}}{B} = \overline{i_n^2} \times 2\tau \tag{7.34}$$

Equating the rms noise to the signal current derivative in (7.21) near $t = 0$ yields

$$\sqrt{\overline{i_n^2}} \frac{\sqrt{2\pi}}{\tau} = i_{S0} \frac{4\pi\sqrt{\overline{\Delta t^2}}}{\tau^2}; \quad \sqrt{\overline{\Delta t^2}} = \frac{\sqrt{\overline{i_n^2}}}{i_{S0}} \frac{\tau}{2\sqrt{2\pi}} = \frac{\tau}{2\sqrt{2\pi}(S/N)_V}$$
$$\sqrt{\overline{\Delta R^2}} = (\Delta R)_{prec} = \frac{c\sqrt{\overline{\Delta t^2}}}{2} = \frac{c\tau}{4\sqrt{2\pi}(S/N)_V} = \frac{(\Delta R)_{res}}{4(S/N)_V} \tag{7.35}$$

and the range precision or uncertainty at high signal-to-noise ratio is given by the range resolution divided by four times the signal-to-noise voltage.

The angular resolution and precision may be calculated in a somewhat similar way by assuming a two-dimensional Gaussian profile image on a split detector such that the current into the separate detector elements is given by

$$i_1 = \frac{2i_S}{\sqrt{2\pi}x_0} \int_{-\infty}^{z} e^{-2x^2/x_0^2} dx; \;\; i_2 = \frac{2i_S}{\sqrt{2\pi}x_0} \int_{z}^{\infty} e^{-2x^2/x_0^2} dx \tag{7.36}$$

where z is the displacement of the image from the center and the e^{-2} power radius is given by x_0 as in Section 3.4. There is, of course, a similar behavior in the y direction, and the full angle measurement may be carried out with a *quadrant array* of four detectors. The currents for the x direction are plotted in Figure 7.14. Taking the derivative of the *difference* of the signal currents with respect to z, we may solve for the rms position error as

$$\frac{\partial i_1}{\partial z} - \frac{\partial i_2}{\partial z} = \frac{4 i_S}{\sqrt{2\pi}\, x_0} e^{-2z^2/x_0^2}; \quad \sqrt{\overline{(\Delta i)^2}} = \frac{4 i_S}{\sqrt{2\pi}\, x_0} \sqrt{\overline{(\Delta z)^2}} = \sqrt{\overline{i_n^2}};$$
$$\therefore \sqrt{\overline{(\Delta z)^2}} = \frac{\sqrt{\pi}\, x_0}{2\sqrt{2}(S/N)_V}; \quad z \ll x_0 \tag{7.37}$$

and taking the angular resolution as the separation of two targets at e^{-1} optical power results in an image spacing on the detector of $z_{res} = x_0\sqrt{2}$,yielding for the angular precision,

$$\sqrt{\overline{(\Delta z)^2}} = \frac{\sqrt{\pi}\, x_0}{2\sqrt{2}(S/N)_V} = \frac{\sqrt{\pi}\, z_{res}}{4(S/N)_V} \tag{7.38}$$

In terms of the far-field beam angles, $\theta_{x,y}$, the angular resolution and precision are given by

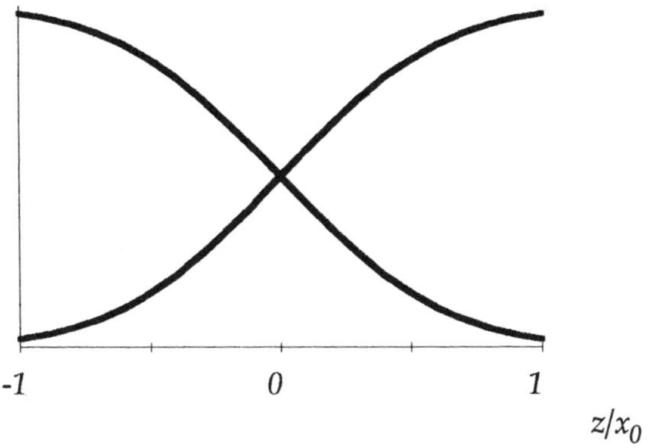

Figure 7.14. Currents from split detector as Gaussian spot of e^{-2} power radius x_o moves a distance z from center.

$$(\theta_{res})_{x,y} = \sqrt{2}\theta_{x0,y0} = \frac{\sqrt{2}x_0, \sqrt{2}y_0}{f}$$
$$\sqrt{\overline{(\Delta\theta)^2_{x,y}}} = \frac{\sqrt{\pi}\theta_{x0,y0}}{2\sqrt{2}(S/N)_V} = \frac{\sqrt{\pi}(\theta_{res})_{x,y}}{4(S/N)_V} \tag{7.39}$$

with f as the focal length of the optical system. Using the error signals from such a split or quadrant detector, closed-loop target tracking can be enabled.

Similar relationships exist for the heterodyne case and these may be found in standard microwave radar texts, such as *(Nathanson, 1990)*. Since a heterodyne radar can also measure frequency, it is also possible to measure the radial velocity using the Doppler shift. The frequency shift is given by $\Delta\nu = 2v/\lambda$, with v the line-of-sight target velocity toward the transmitter. The frequency resolution is approximately $1/\tau$, with τ the pulse width, and the precision, for range and angle as well, is approximately the resolution divided by the *square root* of the $(S/N)_P$, all dependent on the particular pulse or beam shape.

7.4.4 Performance with signal-dependent noise

As in the optical communications case, signal-dependent noise changes the overall behavior of the detection and false-alarm probabilities. We again define the noise current ratio as

$$\delta = \sqrt{\overline{i_{n0}^2}} \,/\, \sqrt{\overline{i_{n1}^2}}$$

as in Eq. (7.20), and rewrite Eqs. (7.28) as

$$p_{fa} = \frac{1}{\sqrt{2\pi \overline{i_{n0}^2}}} \int_{i_T}^{\infty} e^{-i^2/2\overline{i_{n0}^2}} di = 1 - \frac{1}{\sqrt{2\pi}} \int_{-\infty}^{(T/N_0)} e^{-x^2/2} dx$$
$$p_d = \frac{1}{\sqrt{2\pi \overline{i_{n1}^2}}} \int_{i_T}^{\infty} e^{-(i-i_S)^2/2\overline{i_{n1}^2}} di = \frac{1}{\sqrt{2\pi}} \int_{-\infty}^{([S/N_1]-[T/N_1])} e^{-x^2/2} dx \tag{7.40}$$
$$= \frac{1}{\sqrt{2\pi}} \int_{-\infty}^{([S/N_1]-\delta[T/N_0])} e^{-x^2/2} dx; \qquad \frac{T}{N_0} = \frac{i_T}{\sqrt{\overline{i_{n0}^2}}}$$

with $(S/N_1) = (S/N)_V$. We find the detection probability by first finding the threshold current for a given non-signal noise current and false alarm probability. Typical ratios are

p_{fa}	=	10^{-2}	10^{-4}	10^{-6}	10^{-8}
$\frac{T}{N_0} = \frac{i_T}{\sqrt{i_{n0}^2}}$	=	2.33	3.72	4.75	5.61

Then, using the second part of Eq. (7.40) we obtain the results of Figure (7.15) for a false-alarm probability of $p_{fa} = 10^{-6}$. As can be seen, there is a marked reduction in the required signal-to-noise voltage for small values of the noise ratio δ. From Eq. (7.40), the new required $(S/N)_V$ compared with the fixed noise case of Figure 7.10 becomes

$$\left(\frac{S}{N}\right)_{V\delta} = \left(\frac{S}{N}\right)_{V(\delta=1)} - (1-\delta)\left(\frac{T}{N_0}\right) \tag{7.41}$$

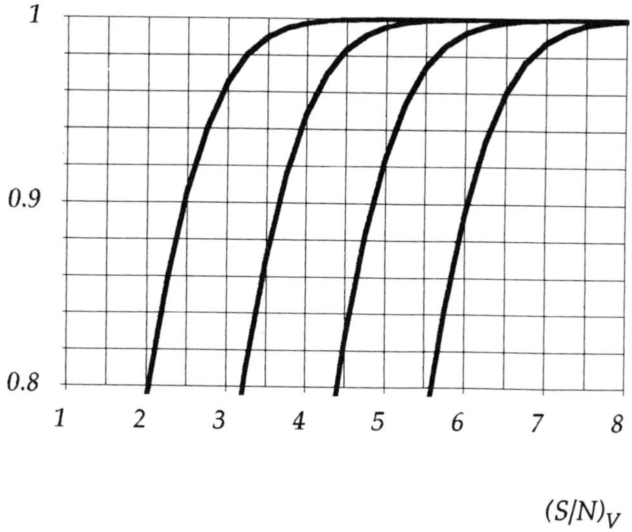

Figure 7.15 Detection probability p_d for several values of noise current ratio δ with a false-alarm probability of 10^{-6}. From left to right, $\delta = 0.25, 0.5, 0.75,$ *and* 1.0.

using (T/N_0) from Eq. (7.28). The reduction is thus most significant at the lowest false-alarm probabilities. Again, as δ approaches zero, the threshold current becomes zero, Gaussian statistics no longer apply, and we must use the more exact Poisson statistics.

Example. A direct-detection radar operates at a wavelength of $1.06\ \mu m$ and utilizes a silicon APD with an ionization ratio of 0.01, a quantum efficiency $\eta = 0.5$, and gain $M = 200$. The bandwidth is $100\ MHz$ corresponding to a $5\text{-}nsec$ pulse, and the APD/amplifier capacitance is $1\ pf$, yielding a required R_{in} of $1600\ \Omega$, and we shall assume a T_N of $50\ K$. Rewriting Eq. (5.20) for an APD, we obtain

$$\left(\frac{S}{N}\right)_V = \frac{P_S}{\sqrt{\frac{2P_S Fh\nu B}{\eta} + \frac{1}{M^2}\left(\frac{h\nu}{\eta q}\right)^2 \frac{4kT_N B}{R_{in}}}} = \frac{\eta P_S}{2Fh\nu B} \frac{1}{\sqrt{\frac{\eta P_S}{2Fh\nu B} + \frac{kT_N}{M^2 q^2 R_{in} B}}} \tag{7.42}$$

with the nonsignal noise being the second term in the denominator. The value of this term for the specified parameters is

$$\frac{kT_N}{M^2 q^2 R_{in} B} = \frac{(kT_N / q)}{M^2 q R_{in} B} = \frac{0.026(50/300)}{(200)^2(1.6 \times 10^{-19})(1600)(10^8)} = 4.2 \tag{7.43}$$

For a detection probability of 90% with a false alarm probability of 10^{-6}, the nonsignal-noise-dependent case would require a $(S/N)_V$ of 6. If we set $\eta P_S / 2Fh\nu B = 15$, then $\delta = \sqrt{4.2}/\sqrt{19.2} = 0.47$, and $(S/N)_V = 15/\sqrt{19.2} = 3.4$, which results in the same performance for the signal-dependent-noise case. Thus, correction for the signal-dependent noise effects results in almost a factor of two reduction in the required transmitter power. Of interest is the number of required photons per pulse at the receiver for this performance This becomes $P_S \tau / h\nu = 15F/\eta = 120$, using $F = 4$ from Figure 5.13.

7.5 Atmospheric Effects

We have not attempted to treat armospheric effects in any detail in our systems analyses other than to include the absorption coefficient α in our performance equations. Figure 7.16 shows the clear weather atmospheric transmission "windows" over paths of the order of a kilometer at sea level as a function of the wavelength. Detailed absorption data as well as a full discussion of atmospheric turbulence may be found in *(Accetta and Schumaker, 1993, Vol. 2)*.

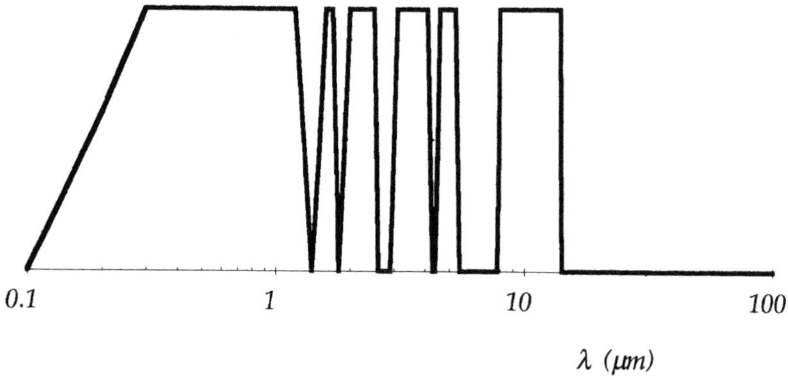

Figure 7.16 Approximate clear weather transmission "'windows" of the atmosphere at sea level.

Problems

7.1 An intrusion alarm uses a HgCdTe photovoltaic mode detector in an optical system with aperture diameter of *2 cm* and focal length of *20 cm*. If an intruding object suddenly fills the detector FOV, we wish to find how small a temperature difference will yield a $(S/N)_V$ of *10*, which would assure a high detection probability and low false-alarm rate. The detector has a wavelength cut-off of $5\ \mu m$, an η of *1*, an area of $10^{-2}\ cm^2$, a zero-bias junction resistance of $10\ \Omega$, and operates at room temperature, *300 K*. Since the detected radiation is on the "tail" of the blackbody curve, it may be approximated by

$$I_{source} = Ke^{-h\nu/kT} = 5.9\ \ watts/m^2\ \ at\ \ 300\,K$$

(a) Find dH/dT in W/m^2K.

(b) What is dP_S/dT for the detector ?

(c) Assume junction noise is dominant and find the *NEP* for a bandwidth of $B = 100\ Hz$.

(d) What is the minimum temperature change that will give a reliable alarm ?

7.2 The *PINFET* amplifier sensitivity for direct detection at the end of Section 7.2 was *-43 dBm* or $5 \times 10^{-8}\ W$.

(a) Find the sensitivity for the same data rate using ideal *heterodyne* performance with $\eta = 0.5$, $B = 90\ MHz$, and the same BER = 10^{-9}.

(b) Find the new achievable range for a fiber attenuation of *0.2 dB/km*, and $P = 1\ mW$

7.3 A museum exhibit is held in a large room with a ceiling height of *10 m*. To supply multilingual narrative, the visitors are supplied with earphones equipped with a PIN-FET /audio amplifier with an optical filter tuned to one of several semiconductor laser wavelengths. In the ceiling, various lasers transmit audio messages in the various languages. All the lasers operate near $0.9\ \mu m$. Each laser radiating area is $1\ (\mu m)^2$, the threshold current is *20 mA*, operating point, *50 mA*, at *15 mW*. The maximum audio signal is thus a sinusoid with a full excursion between

0 and *30 mW*. There are no optics at the transmitter or the detector so that wide-angle transmission and reception is assured.

(a) For a detector area of $2\ mm^2$, find the average or quiescent received optical power assuming the detector is immediately below the laser and pointed in a vertical direction. Use $R = 9\ m$.

(b) To assure reliable operation we wish a large *electrical* audio power signal-to-noise ratio at the amplifier output. For full laser modulation what is this ratio with $R_{in} = 1\ M\Omega$, $T_N = 300\ K$, and $\eta = 0.5$. The audio bandwidth is *10 kHz*.

(c) Estimate the electrical power required to drive the laser transmitter.

7.4 A "cat's eye" retroreflector is shown in the sketch. An incident transmitted wave focuses onto the *diffuse* reflective surface. Part of the reflected power returns to the lens and is propagated back in the direction of incidence as a plane wave. (The retroreflection can occur over a broad range of incidence angles.) Find an expression for the radar cross section σ in terms of A_L, f, and λ. Assume that the lens diameter is small compared to the focal length and that the diffuse reflectivity is unity. *Hint*: See Eq. (1.23).

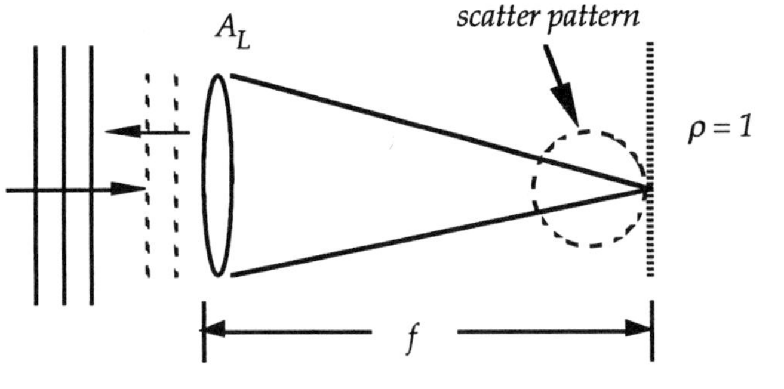

7.5 A photon-counting digital communication system uses on-off coding. If the received background count in each pulse interval has an average value of $n_b = 1$, and the threshold is set at $n_T \geq 4$, find $p[1,0]$, the probability of a "one" being registered when a "zero" is transmitted.

7.6. In the America's Cup race (as seen on TV in 1992), one of the ships was using a pulsed laser rangefinder to measure the range to its competitor. The aperture appeared to be about *10 cm* in diameter and let us assume a range *resolution* of *1.5 m* and a maximum operating range of *1 km*. If the detector requires *1000 photons/pulse* and the laser operates at *1.06 μm*, find the required energy per pulse and peak laser power. Assume the opposing ship is *resolved* by the ranging system and its reflectivity is $\rho = 0.1$.

7.7. You are to design a simple laser radar for automobile speed detection. The radar uses a *30-mW* GaAs laser at *0.9 μm* in the *homodyne ("self-heterodyne")* mode; that is, we transmit continuously and use a small fraction of the transmit laser power as the local oscillator. The Doppler offset, $\Delta\nu = 2v/\lambda$, then determines f_{if}.

(a) Find the Doppler offset for an approaching vehicle at *75 mph (34 m/sec)*.

(b) Assume a beam spot area at the target vehicle of $100\ cm^2$ at *100-m* range and calculate the transmitting and receiving aperture areas.

(c) Find the received power at a range of *100 m*. Use your own engineering estimate of the reflectivity, but state your reasoning.

(d) Because of vibrations and laser short-term stability we need an *i.f.* filter bandwidth of about *10 kHz*. Find the $(S/N)_P$ and the velocity resolution. State any assumptions made in this last calculation.

7.8 A satellite-to-satellite communication system uses a GaAs laser transmitter and operates in the heterodyne mode at a wavelength of $\lambda = 0.9\ \mu m$. The transmitted power is *30 mW*, both A_T and A_R are $3 \times 10^{-2}\ m^2$ and $R = 40{,}000\ km$..

(a) Find the optical signal power at the heterodyne detector.

(b) For a bandwidth, *B*, of *200 MHz*, with $\eta = 0.5$, what is the heterodyne $(S/N)_P$ with *100%* mixing efficiency, neglecting amplifier noise ?

(c) For $R_{in} = 50\Omega$ and $T_N = 300K$, determine the local oscillator power required to obtain this limiting heterodyne performance.

Chapter 8
Systems II. Imaging

To this point, we have considered optical systems that use a single detector element or at most four detectors in a quadrant array for angle measurement. In any event, the detector element or elements are read out individually and then electrically processed. Here we consider imaging systems, where a detector array or continuous detector surface can produce an electrical signal defining a one- or two-dimensional optical image at the focal plane of a receiver or camera. In an array of discrete detectors, the resolution is determined by the spacing of the detectors. In a continuous surface medium such as that of an electron-beam scanned vidicon, the resolution may be stated in terms of the spacing of the *pixels* or picture elements. In this latter case, the resolution or pixel size is not only dependent on the physical structure of the *retina* or scanned surface, but also on the scan pattern and the bandwidth of the output electrical network.

The overall performance of an imaging system may be described by the MTF or *modulation transfer function,* which is determined by the combined performance of the optical system and the imaging detector. The MTF is analogous to the electrical system function *H(f)* but the *f* used for the modulation transfer function is the *spatial* frequency measured in *cycles/unit distance,* usually *cycles/mm* in the image plane or occasionally *cycles/rad* in the far-field object space. The *MTF* or *OTF* (optical transfer function) is treated in many texts on optical systems and is characterized by a *frequency cut-off* or optical *half-power* response at approximately $1/2d$, where d is the half-power diameter of the focal plane image produced by a point source, sometimes called the *point spread function.* The description is completely analogous to the more familiar electric circuit terminology, with the point image width replacing the electrical pulse-width. We do not pursue the MTF characterization in great detail, beyond the obvious relation between the electron-beam scan rate and the output electrical spectrum.

We start first with vacuum image tubes, which utilize either external photo-emission or photoconductivity, then return to single or multiple detector element arrays, and conclude with the most sensitive of the image devices, the *CCD* or charge-coupled device imager. In all of these approaches, the sensitivity is mainly determined by the read-out technique and the associated noise.

8.1 Photoemissive Image Tubes

Vacuum tubes have historically been the most important imaging devices until the recent advances in the CCD. These tubes use either photoemissive or photoconductive area detectors. We first consider photoemissive devices and as a representative example, the *image orthicon,* one of the more popular and versatile forms. In this, as in all image tubes, an electron beam is "raster" scanned over a read-out area as shown in Figure 8.1. In the case of the image orthicon, the scanned surface is that of a semi-insulating sheet as shown in Figure 8.2. The optical image is formed on the negative transparent photocathode and the photoelectrons are accelerated through the mesh striking the read-out plane. Secondary electrons, $\delta \approx 5$, are collected by the slightly positive mesh leaving a positive charge distribution of $(\delta - 1)$ charges per incident electron on the layer, which is a reproduction of the optical image. The scanned electron beam operates at a low enough voltage that the secondary emission ratio is less than one. The beam may then deposit negative charge until the positive charge is cancelled. Any further deposition would cause the target to become more negative than

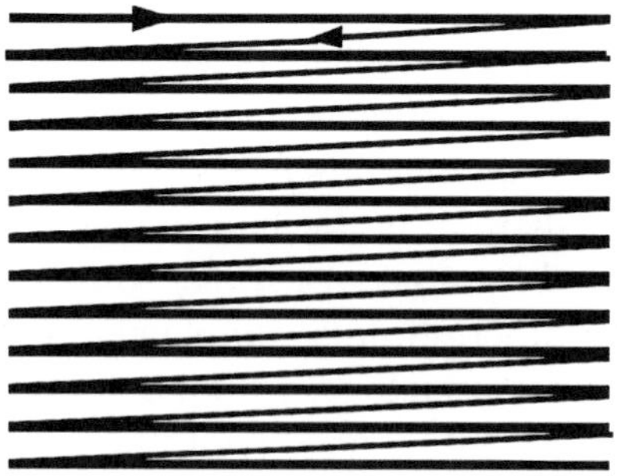

Figure 8.1 Raster scan motion of electron beam. During the thinner return path, the beam moves much faster and the beam current is usually zero.

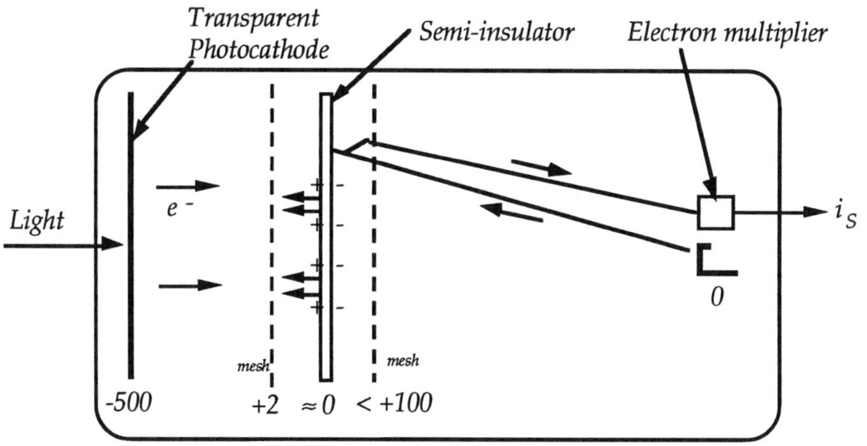

Figure 8.2 Image orthicon, simplified diagram.

the electron-beam cathode and thus repel the beam. The remainder of the beam returns to an electron multiplier similar to that in a photomultiplier tube. The *reduction* in return beam current is then a measure of the incident optical signal on the photocathode or "retina." If we define a "pixel" or picture element width as the product of the electron beam horizontal velocity and the electrical sampling time τ, then one photon in a pixel results in a $(\delta - 1) \approx 4$ *electron* reduction in the return beam current. If the expected peak signal is *40,000 photoelectrons/pixel*, then the incident beam current must be at least *160,000* electrons per sampling time. For low light levels, the input to the electron multiplier thus has an rms fluctuation of $\sqrt{N} = 400$ produced by the shot noise in the electron beam from the thermionic cathode. With this very simple model of the noise, the number of incident photoelectrons required for a $(S/N)_V = 1$ becomes $400/4 = 100$. This quantity is sometimes defined as the noise equivalent electron count or *NEE*. The noise equivalent photon count, $NE\Phi$ is then NEE/η, which for *20%* quantum efficiency becomes *500* for this case. The dynamic range at the specified beam current is then $40{,}000/100 = 400$. A detailed treatment should take into account the noise in the secondary emission process for both the photoelectrons and beam electrons, as well as electron losses at the mesh electrodes. Such calculations and an excellent and extensive review of image tubes can be found in *(Csorba, 1985)*, who defines the *(S/N)* as the *dynamic range*, that is the ratio of the *maximum* signal to the noise. The sensitivity of the image orthicon can obviously be improved by decreasing the beam current with a consequent decrease in dynamic range. We find this to be a common property of most imaging devices. Another important property of this and almost all imaging devices is "frame-to-frame" storage. By this, we mean that the individual pixel or target resolution element integrates the input optical signal over the frame or raster scan time. The signal is then read out and the integration process starts anew.

Higher sensitivity photoemissive tubes *(Csorba, 1985)* operate in similar manner to the orthicon, but replace the semi-insulating target with a target such as silicon where high energy impinging photoelectrons produce hundreds and more internal secondary electrons. The resultant change in local conductance is read out by the vidicon process described later. Another way of amplifying the photoelectron image stream is through the use of a *microchannel plate.* The plate comprises a two-dimensional array of hollow tubes, each with an internal coating that is both resistive and secondary emitting. A high voltage applied from front to back of the array produces an internal longitudinal field and an entering electron is multiplied as it makes successive collisions along its path through the tubular section. Thus if the array density is high enough, the photoelectron image is multiplied with high resolution before striking the target. In either of these modifications, the sensitivity is limited only by noise in the multiplication process rather than by the electron-beam read-out process. The micro-channel plate is also used in *image intensifiers* whose output is a visible image on a phosphor screen rather than an electrically scanned signal.

8.2 Photoconductive Image Tubes

For the longer wavelengths, into the red and beyond, photoemissive image tubes lose sensitivity because of the decrease in available quantum efficiency. This results in poor color rendition in the visible and limits practical applications to wavelengths less than about $1\ \mu m$. The *vidicon* utilizes semiconductor photoconductivity or internal photoemission and thus can operate well into the infrared region of the spectrum, al-

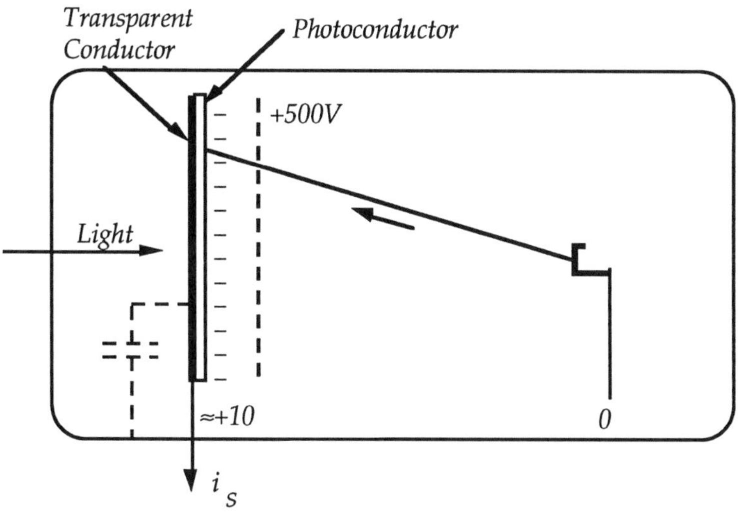

Figure 8.3 Vidicon, simplified diagram.

though cooling is necessary for the longer wavelengths. As shown in Figure 8.3, a raster-scanned electron beam is accelerated to a high voltage for optimum focus and initially charges the surface of the photoconductor to the cathode potential, *0 V*. Each pixel area then loses negative charge to the *10-V* positive transparent conductor at a rate proportional to the light-induced photoconductance produced by the image. The pixel is then recharged every frame time and the resultant signal current, i_S, flows from the conducting plane to an external amplifying circuit. As in the image orthicon, the incident optical signal is integrated over a frame time. Various photoconductors may be used in this device and with cooling, operation can be extended to wavelengths well beyond *1 μm*. Because of the high internal quantum efficiency of semiconductors such as silicon, the color rendition in the visible is extremely uniform. In addition to uniform photoconductive layers, the target may consist of an array of diodes as in the *silicon-target vidicon*. As shown in Figure 8.4, a dense array of silicon *p*-type regions on the surface of an *n*-type layer becomes reverse-biased when charged by the electron beam. Illumination from the left then discharges the diode just as in the photoconduct-

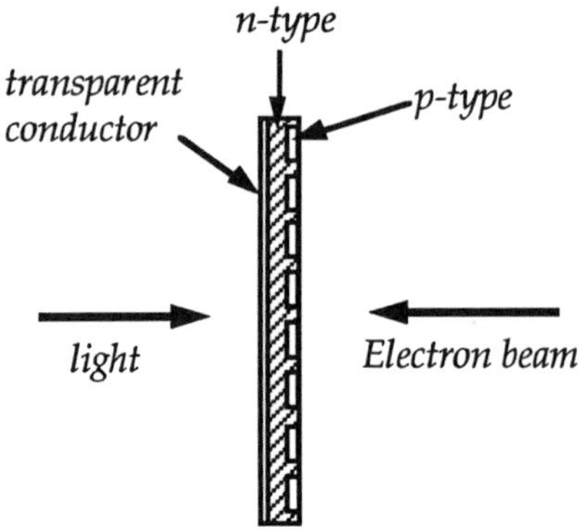

Figure 8.4 Diode array for silicon target vidicon.

or operation. One of the advantages of the diode array is the elimination of frame-to-frame "memory" in the image associated with the finite recombination lifetime in a photoconductor as well as trapping in mid-gap states. This memory or "lag" can produce "smearing" of a moving image.

To treat the sensitivity, we must consider both signal-induced and amplifier noise. The signal-induced noise is the fluctuation in optically induced charge flow during one frame of illumination. Defining N as the number of optically excited photoelectrons in an individual pixel, we then define the number of electrons replaced by the electron beam and flowing in the external circuit as N_S. For the photoconductor, from Eq. (5.4), $N_S = G_pN$, while for the diode array, $N_S = N$. We note that shot noise in the electron beam does not contribute in this case, since the number of electrons replaced at each pixel is determined by the requirement to recharge the surface to $V = 0$, and thus is not a function of the arrival rate. We now calculate the fluctuation in N_S for the two cases as

$$\begin{aligned} \overline{\Delta N_S^2} &= 2G_p^2 N; \quad \textit{photoconductor} \\ \overline{\Delta N_S^2} &= N; \quad \textit{diode array} \end{aligned} \tag{8.1}$$

the first expression derived from Eq. (5.5), using $B = 1/2\tau$, with τ as the frame time. We may now write the signal and noise currents in the external circuit, including the amplifier noise, as

$$i_S = \frac{qN_S}{\tau_S} = 2qN_S B; \quad \overline{i_n^2} = \frac{q^2 \overline{\Delta N_S^2}}{\tau_S^2} + \frac{4kT_N B}{R_{in}} = 4q^2 \overline{\Delta N_S^2} B^2 + \frac{4kT_N B}{R_{in}}$$

$$i_S = 2qG_p NB; \quad \overline{i_n^2} = 8q^2 G_p^2 NB^2 + \frac{4kT_N B}{R_{in}}; \quad \underline{\textit{photoconductor}} \tag{8.2}$$

$$i_S = 2qNB; \quad \overline{i_n^2} = 4q^2 NB^2 + \frac{4kT_N B}{R_{in}}; \quad \underline{\textit{diode array}}$$

with τ_S being the sampling time given by $1/2B$, where B is the bandwidth of the amplifier circuit. Ignoring the amplifier noise and setting the signal and noise currents equal yields the respective $NEEs$ of $\sqrt{2}$ and 1, the latter, the photo-electron counting limit. In the amplifier-noise-limited case, the NEE becomes

$$NEE = \frac{1}{G_p}\sqrt{\frac{kT_N}{q^2 R_{in} B}} = \frac{1}{G_p}\sqrt{\frac{2\pi k T_N C_{in}}{q^2}} \tag{8.3}$$

with $G_p = 1$ for the diode array and C_{in} the total shunt capacitance for the tube output and the amplifier. The tube capacitance is *not* the front-to-back capacitance of the photoconductive layer or the diode array, but is the capacitance of the target to ground as indicated in Figure 8.3. The photoconductive gain $G_p = \tau_r/\tau_t$, the ratio of the carrier recombination time to the transit time, is generally of the order of unity or less for most photoconductors. Otherwise a long recombination time would allow diffusion to neighboring pixels, thus reducing the resolution. Typical resolutions are 10 to $30\ \mu m$ with a commensurate active layer thickness and p-type region spacing in the silicon target vidicon.

Example: A vidicon used for television operates at a bandwidth of $4\ MHz$ and the tube plus amplifier capacitance is $5\ pf$. Assuming an amplifier noise temperature of $T_N = 100K$, the noise equivalent electron count becomes for a diode array

$$NEE = \sqrt{\frac{2\pi k T_N C_{in}}{q^2}} = \sqrt{\frac{2\pi(.026)(100/300)(5\times 10^{-12})}{1.6\times 10^{-19}}} = 1650 \quad (8.4)$$

so that the system is obviously amplifier limited. With a quantum efficiency of *50%*, the noise equivalent photon count is $N\,E\Phi$ = *3300*.

The noise equivalent photon count for each pixel decreases with decreasing bandwidth or read-out rate. In addition, the sensitivity to a given irradiance at the retina increases with frame time, since both photoemissive and photoconductive image tubes integrate the photoelectron stream. The maximum frame time for a television camera is limited by frame-to-frame "flicker". Longer frame times can be used for other applications such as astronomy; however, the dark current of the photoconductor or silicon diode then becomes the limiting factor. Finally, the sensitivity to a given irradiance increases with increasing pixel size, but with a consequent loss in resolution. As a reference point for typical camera tube irradiances, let us consider the image of a perfectly white diffuse surface illuminated at normal incidence by the unattenuated sun. We shall assume an optical system of $f/\# = 2$, a pixel size of *20 μm* by *20 μm*, wavelengths of *1 μm* and shorter. The irradiance at the retina becomes

$$I_i = \frac{H_S}{(2f/\#)^2} = \frac{1400(I > 2.47)}{16} = 60 \;\; W/m^2 \qquad (8.5)$$

using Figure 1.14 with $h\nu = 1.24\ eV$ and $T = 5800\ K$. Approximating the photon energy as that at *1 μm*, the rate of photon arrival per unit area becomes

$$r/A = (P/A)/h\nu = 60/(1.24)(1.6\times 10^{-19}) = 3\times 10^{20}\, photons/m^2 sec \qquad (8.6)$$

and for a a pixel area of *400* μm^2 a flux of 1.2×10^{11} *photons/sec*. For a *1/60-sec* frame time (TV frame rates are *60* interlaced frames per second resulting in *30* full frames per second), there will be 2×10^9 *photons/pix-*

el. Allowing for atmospheric attenuation, non-normal solar incidence, and a reasonable range of reflectances in an image reduces the average expected flux to approximately 10^8 *photons/pixel* for a "sunny" day image. Light from a full moon, from the example in section 1.4, would result in an average of about 10^3 *photons/pixel*, below the noise level of the vidicon, and only a factor of two higher than the image orthicon discussed in the previous section.

The sensitivity of the vidicon may be improved by using a return beam read-out similar to that of the image orthicon. An even more sensitive mode of operation for both the orthicon and the vidicon is the *isocon* return beam technique. During the initial time when electrons are being replaced on the target surface by the electron beam, a fraction are reemitted as secondaries and *scattered* over a reasonably large angle. Once the correct negative charge is reestablished, the remaining beam electrons are *reflected* in a narrow beam angle. By collecting and multiplying *only* the scattered electrons, the excess main beam shot noise is therefore eliminated. We have not treated the important topics of retina uniformity, beam focus and deflection techniques, and vidicon dynamic range. These can be found in *(Csorba, 1985)*.

8.3 Detector Arrays

Discrete detectors may be used to form an image by scanning of a single element or a linear array as shown in Figure 8.5.(a) and (b). The detectors themselves are not scanned, rather, the image field is scanned over the element or array using a rotating cylindrical mirror assembly or assemblies in the optical path. The techniques are discussed in detail in *(Accetta and Shumaker, 1993)*. A full two-dimensional array may be used

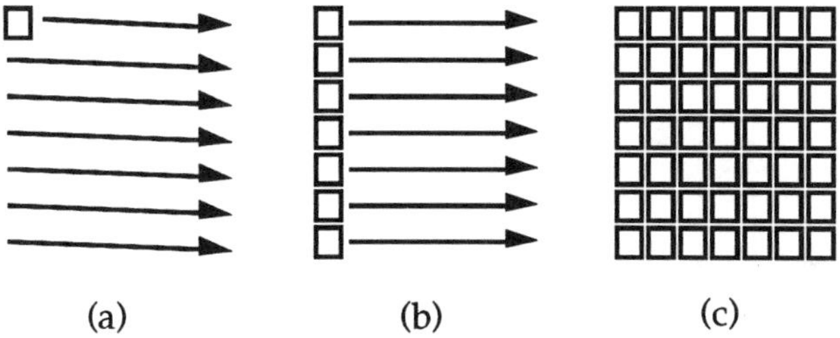

Figure 8.5 (a) Single element , (b) linear array scan and (c) area array.

as shown in Figure 8.5(c). The single detector system uses standard techniques for read-out, with the amplifier bandwidth determined by the scan rate. Linear and area arrays may also be read out in parallel with individual amplifiers and data recording for each pixel. Alternatively, the parallel outputs from the amplifiers can be integrated, stored, and sequentially sampled to produce a serial data stream. In either case, detector uniformity may seriously limit image quality and data reliability. This problem can be overcome by calibration and the resultant stored gain and bias adjustments for individual elements.

For large arrays, the complexity of individual detector amplifiers may be overcome by the use of capacitive storage and sequential readout of the detectors. The sampling shown schematically in Figure 8.6 may be accomplished by a shift register-driven set of FET switches, but the added capacitance of the output line increases the value of *NEE*. For background-limited operation at long wavelengths, this is not a serious problem; however, at near-infrared and visible wavelengths, the amplifier noise becomes limiting. From Section 4.5, the *NEE* is given by

$$NEE = \frac{C}{q}\sqrt{\frac{4\gamma kTB}{g_m}} \tag{8.7}$$

which, for a read-out line capacitance of *10 pf*, a g_m of *5 mS*, $\gamma = 1$, and a bandwidth of *1 MHz*, becomes *250*, if we use correlated double sampling, that is, measure the voltage immediately before and after each individual switch is closed, as discussed in Section 4.5. Otherwise, *kTC* noise results in a noise equivalent electron count given by

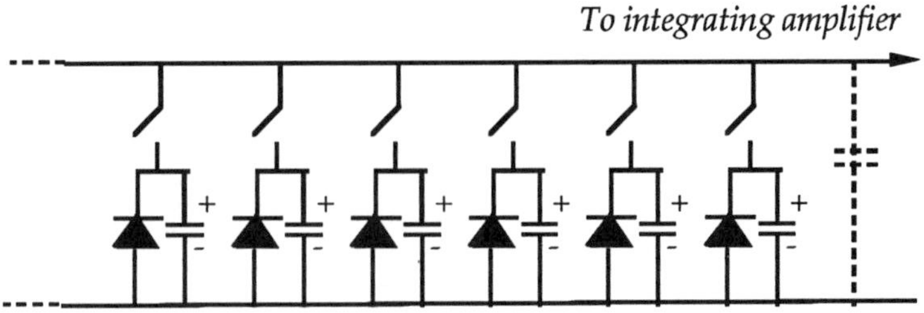

Figure 8.6 Schematic representation of array sequential read-out

$$NEE = \frac{\sqrt{kTC}}{q} = \sqrt{\frac{(kT/q)C}{q}} = \sqrt{\frac{0.026(10^{-11})}{1.6 \times 10^{-19}}} = 1270 \qquad (8.8)$$

The dynamic range of the detectors is determined by the storage capacitance at each diode, consisting of the detector internal capacitance plus, if desired, an external capacitor. With a net capacitance of $1\ pf$ and a maximum voltage swing of $5\ V$, the net electron storage becomes $N = CV/q = 3 \times 10^7$ *electrons*, or a dynamic range of more than 10^4. If the same array is operated at long wavelengths and is background-limited, then the background electron count might be set at 10^7 *electrons*, with a resultant $(NEE)_{BL}$ of 3.23×10^3 and a net NEE of 3.4×10^3 including the read-out contribution of Eq. (8.8). It is in this latter mode of operation that array uniformity becomes critical and it is often necessary to post-process the data using individual detector calibration data. Two-dimensional array readout is accomplished by sequential readout of each row of detectors in the array.

8.4 Charge-Coupled Device (CCD) Arrays

The read-out techniques discussed in the preceding sections can be performed using an integrated detector and read-out array in a semiconductor chip. This technique is particularly attractive for silicon detectors because of the large number of elements and uniformity available with current VLSI technology. Various integrated detection and read-out techniques are discussed in *(Waynant and Ediger, 1994, Ch. 18)* and *(Accetta and Shumaker, 1993, Vol. 3)* and the most sensitive and frequently used is the charge-coupled device or CCD.

Figure 8.7 is the energy diagram for a p-type silicon metal-oxide semiconductor (MOS) structure with two different positive voltages on the metal electrode. The heavy lines have the highest positive potential and therefore attract electrons into the well adjacent to the oxide layer. These electrons are produced by thermal generation (dark current), by photoionization (photocurrent), or by transfer from an adjacent region (dimension into the paper) at a lower positive voltage, as shown by the light-lined diagram. An alternative representation is in Figure 8.8, which shows the phase diagram for the electrode voltage drive, a short

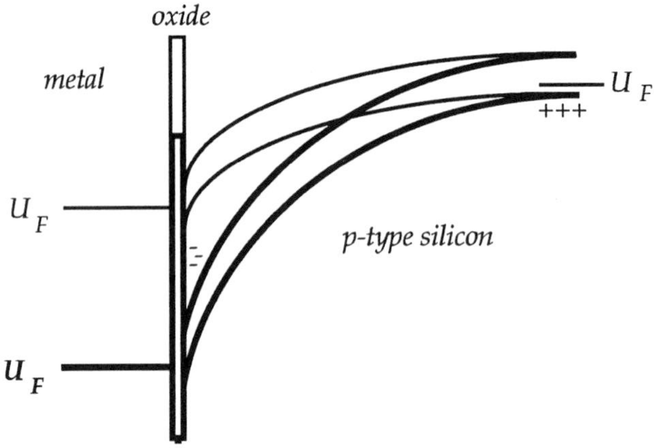

Figure 8.7 Energy diagram for two adjacent *CCD* wells. The darker lines show the more positive metal electrode

cross-section of a*CCD* structure, and the potential changes as the electron charge packet is transferred to an adjacent electrode. In an imaging mode, the incident light produces electrons that travel to the nearest deep well and, after a frame time, the charges are "clocked" one pixel or three electrode units to the right, and in the two-dimensional structure of

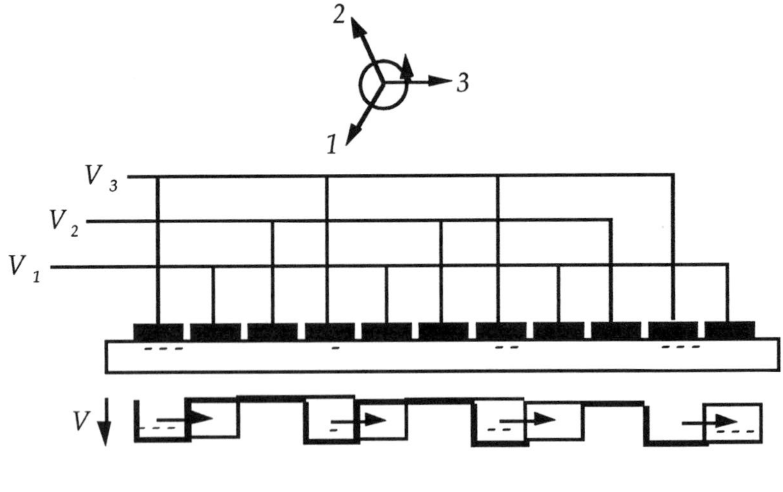

Figure 8.8 Phase diagram for electrode voltage drive, *CCD* cross-section, and associated potentials

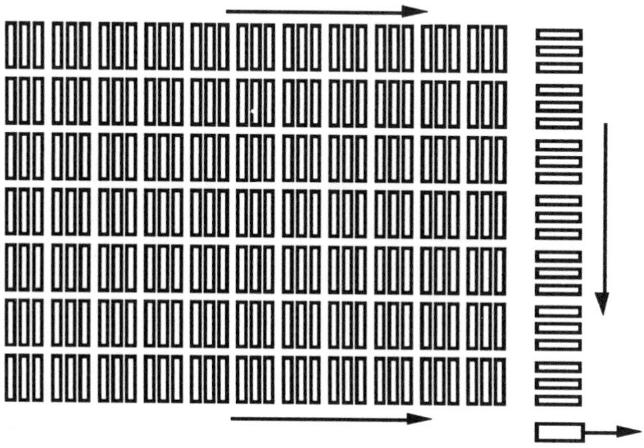

Figure 8.9 CCD imaging array showing read-out sequence

Figure 8.9, stored in a column register of CCD wells. The column is then read out to a charge-sensing electrode and then the sequence repeated until the full frame has been read out. The illumination may fall on the back of the thin silicon substrate or on the surface if the metal electrodes are transparent. Since the readout takes a full frame time, many CCD imagers are twice the image size and a full frame is transferred to an unilluminated region and then read out while the next frame is being imaged. The key to successful operation is a high transfer efficiency from well to well, that is, negligible trapping or loss of electrons. This is critically dependent on the oxide-silicon interface in the "surface-channel" device we discuss here. It is also possible to build "buried-channel" structures where the well is isolated from the oxide by a potential barrier *(Howes and Morgan, 1979)*. In addition, two-phase rather than three-phase operation may be accomplished by varying the impurity doping along each gate electrode.

The sensitivity of the CCD can be extremely high because of the single charge sensor electrode and the resultant opportunity for a low capacitance. The output circuit, as shown in Figure 8.10, is a resetting integrator with an FET switch and an FET input stage amplifier. Both FETs are integrated on-chip thus minimizing the capacitance. The sensor electrode is initially set to $+V$ with the switch closed, then the switch

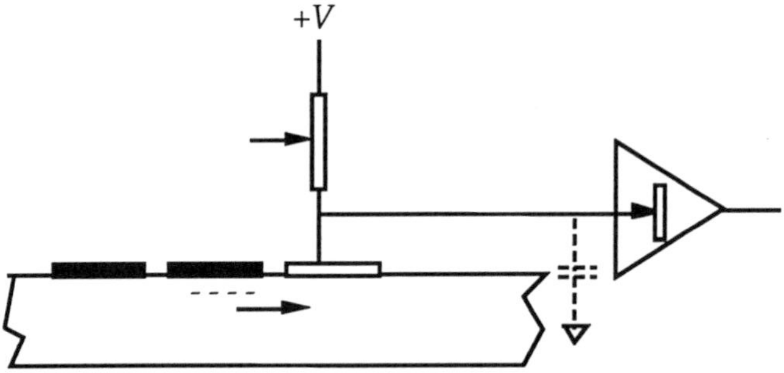

Figure 8.10 Output charge-sensing circuit.

is opened and the amplifier output voltage is sampled. Then the packet of charge is transferred to the sensor electrode and the voltage sampled again. This *correlated double sampling* eliminates the *kTC* noise and the resultant *NEE* is given by Eq. (8.7). As an example, with a net electrode-amplifier capacitance of *0.1 pf,* an FET transconductance of *0.5 mS* with $\gamma = 1$, and a *TV*-compatible bandwidth of *4 MHz*, the *NEE* becomes *7*. With appropriate design and lower bandwidths or readout rates, it is possible to reduce the NEE of the readout stage to less than one, leading to single photo-electron count capability. Such performance depends upon reducing dark current by cooling, negligible trapping in individual wells, and a reduced dynamic range as determined by the capacitance of the readout stage.

The CCD array is the key element in the "camcorder" or video camera as well as in astronomy where it now competes successfully with photographic plates. In the camera application, three-color operation can be accomplished by dividing the input image beam spectrally and utilizing three registered CCD arrays. Most portable cameras, however, use color-stripe filters over three adjacent rows and each three rows supplies the three-color output for one scan line. Finally, it should be noted that the fixed geometrical layout of the CCD pixels offers excellent metric and rectilinear performance compared to the vacuum image tube, where the electron-beam scanning can produce distortion as well as temporal changes in position and magnification due to voltage changes and stray electromagnetic fields.

Appendix A: Answers to Problems

1.1 (a) 3.2×10^{-17}
(b) $r = 7.9 \times 10^{4}$ photons/sec

1.2 $du = \dfrac{2h\nu d\nu}{c(e^{h\nu/kT} - 1)} \Rightarrow \dfrac{2kTd\nu}{c}$

1.3 $P = 128$ W; $r = 6.45 \times 10^{21}$ photons/sec

1.4 $\varepsilon = 0.49$

1.5 $dT = 7.5$ K

1.6 $T(mercury) = 483$ K
$T(Pluto) = 47.7$ K

1.7 $G_H = 6.13$ W/m^{2}K

1.8 $T = 820$ K

1.9 $(1/P)(dP/dT) = 0.016$ K^{-1}

1.10 $N = 2.1 \times 10^{11}$ photons

1.11 $NE\Delta T = 1.3 \times 10^{-4}$ K

1.12 $N/A = 8.6 \times 10^{10}$ photons/cm^2

1.13 Q.E.D.

1.14 Tungsten, 2.9 %
Halogen, 5 %

1.15 $T = 904$ K

1.16 $r = 1.1 \times 10^{12}$ photons / sec

2.1 Q.E.D.

2.2 (a) 2.3 A, 86 GHz
(b) $\gamma_t = 23.4$ cm^{-1}

2.3 $\lambda_{min} = 69$ μm

3.1 $\lambda = 76$ A, $a = 4.6$ A

3.2 $\eta_{ext} = [\lambda(\mu m)/1.24]P/I$

3.3 $T = 18540$ K

3.4 (a) $D = 4.9$ km
(b) $D = 3$ μm

4.1 (a) $\overline{i_n^2} = 3.9 \times 10^{-17}$ A^2
(b) $i_S = 1.9 \times 10^{-6} \Delta T$ A
(c) $NE\Delta T = 3.3 \times 10^{-3}$ K

4.2 $\overline{i_n^2} = 1 \times 10^{-18}$ A^2
$NE\Delta T = 3.3 \times 10^{-3}$ K

4.3 Q.E.D.

4.4 $\Re = \dfrac{\eta\lambda(\mu m)}{1.24}$ A/W

4.5 (a) $R_{in} = 231\ \Omega$, $T_N = 22$ K
(b) $(i_n)_{\sqrt{f}} = 2.3$ pA/$\sqrt{Hz}$
(c) $NEP = 2.85 \times 10^{-10}$ W

5.1 $NEP = 3.3 \times 10^{-7}$ W

5.2 $\eta = 0.0014$

5.3 (a) $\eta = (1 - e^{-\alpha d})$
(b) $d = 0.693/\alpha$

5.4 $(NEP)_{SL} = 2\Gamma h\nu B/\eta$
$(NEP)_{DL} = \dfrac{h\nu}{\eta q}\sqrt{2q\Gamma i_D B}$

5.5 $G = 1.14 \times 10^{5}$

5.6 $T = 253$ K

5.7 (a) $R_{in} = 231\ \Omega$, $T_N = 84.8$ K
(b) $(i_n)_{\sqrt{f}} = 4.5 \times 10^{-12}$ A/Hz$^{1/2}$
(c) $(NEP)_{AL} = 5.6 \times 10^{-10}$ W

5.8 (a) $I_s = 6.7 \times 10^{-5}$ A
(b) $(NEP)_{AL} = 1.1 \times 10^{-9}$ W
(c) $P_{max}/NEP = 7500$

5.9 (a) $\overline{i_n^2} = 2qFM^2 i_{S0}B + \frac{4kT_NB}{R_{in}}$

(b) $i_{S0} = qFB + \sqrt{(qFB)^2 + \frac{4kT_NB}{M^2R_{in}}}$

(c) $NEP =$

$$2.65 \times 10^{-10}\left(M + \sqrt{M^2 + \frac{1.3 \times 10^7}{M^2}}\right)$$

(d) $(NEP)_{min} = 3.7 \times 10^{-8}$ W

6.1 (a) $\frac{P_{if}}{P_S} = 2\left(\frac{q}{h\nu}\right)^2 P_{LO}R$

(b) $\frac{P_{if}}{P_S} = 13$

(c) $(\overline{i_n^2})_{shot} = 2.6 \times 10^{-13}\ A^2$

(d) $(\overline{i_n^2})_{amp} = 1.7 \times 10^{-15}\ A^2$

6.2 Q.E.D.

6.3 $\left(\frac{S}{N}\right)_V = \frac{P_S}{2\mu h\nu B\sqrt{N + \frac{P_S}{\mu h\nu B}}}$

7.1 (a) $dH/dT = 0.19$ W/m²K

(b) $dP_S/dT = 4.8 \times 10^{-10}$ W/K

(c) $NEP = 1 \times 10^{-10}$ W

(d) $\Delta T = 2.1$ K

7.2 (a) Sensitivity = – 59.2 dBm

(b) $R = 296$ km

7.3 (a) $P_S = 4.6 \times 10^{-10}$ W

(b) $NEP = 3.6 \times 10^{-11}$

(c) $P_{DC} = 69$ mW

7.4 $\sigma = 4A_L^3 / f^2\lambda^2$

7.5 $p[1,0] = 0.019$

7.6 $U = 7.5 \times 10^{-7}$ J, $P = 60$ W

7.7 (a) $\Delta\nu = 76$ MHz

(b) $A = 0.81$ mm²

(c) $P_S = 7.7 \times 10^{-14}$ W, $\rho = 0.01$

(d) $(S/N)_P = 28 = 14.5$ dB

7.8 (a) $P_S = 2.1 \times 10^{-8}$ W

(b) $(S/N)_P = 240$

(c) $P_{LO} >> 2.9$ mW

References

Accetta, J. S., and Shumaker, D. L., eds. (1993). *The Infrared and Electro-Optical Systems Handbook,* Environmental Research Insitute of Michigan, Ann Arbor, Michigan.

Agrawal, G. P., and Dutta, N. K. (1993). *Semiconductor Lasers,* 2nd ed. Van Nostrand Reinhold, New York.

Bhattacharya, P. (1994). *Semiconductor Optoelectronic Devices.* Prentice Hall , Englewood Cliffs, NJ.

Bloembergen, N. (1956). *Phys. Rev.,* 104, 324.

Boyd, G. D., and Gordon, J. P. (1961). *Bell Syst. Tech. J.,* 40, 489.

Boyd, R.W., (1983). *Radiometry and the Detection of Optical Radiation.* John Wiley and Sons, New York.

Callaway, J. (1991). *Quantum Theory of the Solid State,* 2nd ed. Academic Press, San Diego.

Csorba, I. P. (1985). *Image Tubes.* Howard W. Sams & Co., Indianapolis.

Davenport, W. B., and Root, W. L. (1958). *Random Signals and Noise,* McGraw-Hill, New York.

Einstein, A. (1917). *Phys. Zeit.,* 18, 121.

Emmons, R. B. (1967), *J. Appl. Phys.,*38, 3705.

Goodman, J. W. (1985). *Statistical Optics.* John Wiley and Sons, New York.

Gordon, J. P., Zeiger, H. J., and Townes, C. H. (1955). *Phys. Rev.,* 99, 1264.

Gowar, J. (1984). *Optical Communication Systems.* Prentice Hall, Englewood Cliffs.

Howes, M. J., and Morgan, D. V., eds. (1979). *Charge-coupled Devices and Systems.* John Wiley and Sons, New York.

Iga, K., and Koyama, F. (1993). In *Surface Emitting Semiconductor Lasers and Arrays.* (G. A. Evans and J. M. Hammer, eds.), pp. 71-118. Academic Press, San Diego.

Kingston, R. H. (1978). *Detection of Optical and Infrared Radiation.* Springer, New York.

Maiman, T. H., (1960). *Nature,* 187, 493.

McIntyre, R. J. (1966). *IEEE Trans. Electron Dev.,* ED-13, 164.

Nathanson, F.E., (1990). *Radar Design Principles.* McGraw-Hill, New York.

Ramo, S., Whinnery, J. R., and Van Duzer, T. (1984). *Fields and Waves in Communication Electronics,* 2nd ed. John Wiley and Sons, New York.

Reif, F., (1965).*Fundamentals of Statistical and Thermal Physics.* McGraw-Hill, New York.

Rice, S.O.,(1954). In *Selected Papers on Noise and Stochastic Processes* (N. Wax,ed.). Dover, New York.
Ross, A. H. M. (1970). *Proc. IEEE*, 58, 1766.
Saleh, B. E. A., and Teich, M. C. (1991). *Fundamentals of Photonics*. Wiley-Interscience, New York.
Schawlow, A. L., and Townes, C. H., (1958). *Phys. Rev.*, 112, 1940.
Siegman, A. E. (1966). *Proc. IEEE*, 54, 1350.
van der Ziel, A. (1970). *Noise: Sources, Characterization, Measurement*. Prentice Hall, Englewood Cliffs.
Wang, S. (1989). *Fundamentals of Semiconductor Theory and Device Physics*. Prentice Hall, Englewood Cliffs, NJ.
Waynant, R. W., and Ediger, M. N., eds. (1994). *Electro-optics Handbook*. McGraw-Hill, New York.
Wolfe, W. L., and Zissis, G. J., eds. (1985). *The Infrared Handbook*. Environmental Research Institute of Michigan, Ann Arbor,Michigan.
Wroblewski, R. (1988). In *GaAs MESFET Circuit Design* (R. Soares, ed.). Artech House, Norwood, MA.
Yariv, A. (1991). *Optical Electronics*, 4th ed. Saunders/Holt, Rinehart and Winston, Philadelphia.
Zory, P. S., Jr. (1993). *Quantum Well Lasers*. Academic Press, San Diego.

Index